I WAS BLIND BUT NOW I SEE

EVOLUTION - CREATION

MARILYN OAKLEY

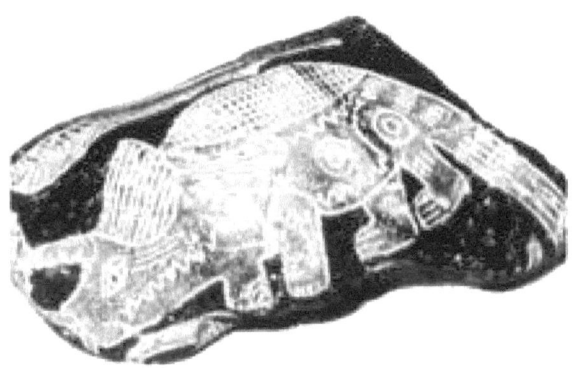

Dinosaur drawn from the Ica period on a stone. This is the dinosaur we know today as the Triceratops. Is it possible that this was made by someone who actually saw a dinosaur?

Unless otherwise stated, any verses, scriptures are from the King James Version of the Holy Bible. Also, any scriptures contained in any articles I have used from other sources, used their own.

© 2007 Marilyn Oakley
ISBN: 978-0-6151-5160-1

First printing

All rights reserved. No part of this book may be reproduced, stored in a retrieval system or transmitted in any form or by any means without the prior written permission of the author, except by a reviewer who may quote brief passages in a review to be printed in a newspaper, magazine or journal.

I'd like to dedicate this book to:

My beloved daughter Angela (Angie) L. Fuerstenberg
February 24, 1980 - February 24, 2005

John 10:27-28 "My sheep listen to my voice; I know them, and they follow me. I give them eternal life, and they shall never perish; no one can snatch them out of my hand."

May she be doing all the things in Heaven that she could not do down here on Earth.

A Healing Prayer:
Dear heavenly Father,
You intend my body to be a temple of the Holy Spirit. May it be your gracious will that I enjoy your healing power, that I may seek you, serve you, enjoy you and depend on you, through the physical life you have given me, in Jesus Christ, my Lord.
Amen

Written in Remembrance of my darling daughter.
© Marilyn Oakley
I'll vision as I close my eyes
Your smiling face, peaceful heart
Lifting beyond the deep blue sky
Dancing past the brightest star

A voice softly whispers in the night
Says hush my heart, don't be sad
She will celebrate her new life
As she goes forth holding Jesus' hand

So long my darling and …
Spread your beauty - Give it new birth
Find happiness - Beyond this earth
Arise precious one - Now journey far
Someday I'll fly - To where you are
~~~
You had a smile and love for everyone
You touched so many you became their sun
You brightened their world even when they were sad
You held their heart and you held their hand
You gave so much in such little time
Your precious heart forever mine
Going places not yet gone before
Holding God's love forevermore

"Scientists have no proof that life was not the result of an act of creation." Robert Jastrow, The Enchanted Loom: Mind In The Universe (1981), p. 19 [a leading astronomer]

Paul Davies a world-renowned scientist makes this statement of the First cell – how? "Nobody knows how a mixture of lifeless chemicals spontaneously organized themselves into the first living cell." (The Australian Centre for Astrobiology Macquarie University Sydney, in Scientist July 12, 2003 179|2403|:32)

# Table Of Contents

| | |
|---|---|
| Prologue | Page 1 |
| Evidence for Evolution | Page 4 |
| The Big Bang | Page 6 |
| Catastrophism | Page 7 |
| Fossil Record | Page 9 |
| Complexity of Design | Page 12 |
| Famous Evolutionists Quotes | Page 17 |
| Cross Breeding & Mutations | Page 19 |
| Creation/Intelligent Design | Page 20 |
| Dinosaurs in the Bible | Page 23 |
| The Flood | Page 26 |
| Eve Created from Adam | Page 27 |
| Cain's Wife | Page 28 |
| Dead Sea Scrolls | Page 29 |
| Micro-Macro Evolution | Page 29 |
| Survival of the Fakest | Page 31 |
| Human Culture | Page 34 |
| Utterly Impossible | Page 35 |
| Humans Formed from Dust | Page 40 |
| Haeckel's Embryos | Page 44 |
| In Conclusion | Page 46 |
| Resources & References | Page 54 |
| Recent News (2007) | Page 55 |
| Our Emotion Our Minds | Page 56 |
| Charles Darwin Bio | Page 57 |
| History Of Evolution | Page 58 |
| Is Atheism Against The Law? | Page 62 |

# I Was Blind But Now I See    Evolution - Creation

## Prologue

What provoked me to write this book? I wanted to know how evolution fit into saying we evolved over what the Bible said of God creating all. I've wondered this for many years. So, I finally decided to search for answers and found what I'm going to write about.

I'd like to thank those who granted me permission to obtain information from them. They are stated within my book. At the end of my book are some great web links to explore.

I've researched before compiling this book so the facts I've written are to the best of my knowledge. "I don't think evolution is right and in my book I'm going to show the reasons why I think their theory is wrong and how Creation gives a better account."

The theory of evolution was first brought to the attention of the scientific world in the nineteenth century. It was expressed by the French biologist Jean-Baptiste Lamarck, in his *Zoological Philosophy* (1809). He thought that all living things were capable with a vital force that impelled them to evolve toward greater complexity. He also conceived that organisms could pass on to their offspring traits acquired during their lifetimes.

This evolutionary model of Lamarck's was without any foundation by the discovery of the laws of genetic inheritance. In the middle of the twentieth century, the discovery of the structure of DNA revealed that the nuclei of the cells of living organisms possess very special genetic information. This special genetic information could not be altered by "acquired traits."

The evolutionary theory was formulated by another natural scientist. This natural scientist was Charles Robert Darwin, and the theory he formulated is known as "Darwinism." He based his theory on various observations he made on a five-year voyage around the world. What made his theory different than Lamarck, Darwin emphasized on "natural selection." He speculated that there is a struggle for survival and through natural selection resulted in the survival of a stronger species.

Charles Darwin developed his theory when science was still in a primitive state. Under primitive microscopes, life appeared to have a very simple structure. This error formed the basis of Darwinism. The primitive understanding of science in his time rested on the assumption that living things had very simple structures.

Of course, evolution will tell you they have many facts, proof and evidence of their theory. In my book, I will present those facts and or theories. Evolutionists will also claim the Creation is not a science, but actually it is a science and holds many facts as well.

Spontaneous generation is the belief that life came from non-living matter on how life began according to the evolutionist. Creationists believe that life was created by an intelligent supernatural being (God).

The law of biogenesis consists of two parts. With the first part stating that living things only come from other living things and not from non-living (dead) things. Life only comes from life. The second part states that when living things procreate, their offspring are the same type of organism they are. Simply put, sharks come from sharks, snakes from other snakes and orange trees from orange trees, etc. All living organisms alive today are a product of and evidence of biogenesis.

Evolution depends on current scientific theories to show the origin of man and the universe. The problem with this is science is constantly changing. Newer and different theories are always being formed concerning our origin. So when one theory is found to be false, another theory is quickly claimed to cover the first error. For example concentric theory taught that sun and planets revolved around the earth in $15^{th}$ century. And in the 17 – $18^{th}$ century, the Phlogiston theory taught that every substance that burns contains a mysterious ingredient called "Phlogiston", which was later shown to be oxygen. It was

## Marilyn Oakley

even once accepted as a scientific fact that mice came from wheat. Evolutionists went hunting for proof when evolution was offered as fact being built on merely hypothesis.

Evolution, it is still printed in our public and college textbooks because many of our "modern" scientists are simply bound and determined never to admit that there is another theory to consider of how we got here.

Why would I have not believed Evolution? Science class presented it as a fact. Today, millions of students are still taught that we are a byproduct of evolved life forms of primordial ooze. Until time and chance brought forth, finally, humanity over millions of years. And in respecting my elders, who were authoritative figures, why would they teach anything that wasn't true?

If evolution provides statements that are only speculated, then perhaps two types of history/science classes should be taught in our public schools, one for evolution and one for Creation/Intelligent Design. Then the parents and or students can decide which one they'd like to take. Or even offer both as options.

Yes, everyone has to the right to believe or not in the teachings of the Bible, but ethnically, creation science should be introduced into the school system as well, so that a child does have a choice. If only one is introduced and not the other, how can that child choose then? Is he/she not going to be smart enough as they grow older to choose what he or she wants to believe? It doesn't have to be up to a parent to teach the Bible or creation. Not all parents believe or even know, and the option of creation science should also be offered in public schools as well. Then the child can decide as an individual what he'd/she'd like to believe, even if their parents disagree with their decision, but it is his or hers to decide.

In my opinion, it seems we have a bit of communism going on, when someone obviously has the authority to say what should be taught in schools and what should not be taught in schools. Evolutionists can say creation is only a theory, but evolution is only a theory as well. Evolution should be honest enough to state they don't have facts, proof or evidence to back up any claims they make. That particles-to-people evolution is an unsubstantiated *hypothesis* or *conjecture*.

America is supposed to be freedom of speech, freedom of choice, how do we have a choice or speech, when it isn't offered? Who chooses to only teach evolution science over creation science?

In my opinion, evolution can give the wrong impression, or those that take it literally. Evolution teaches survival of the fittest, where the strong take over the weak, and this taken in the wrong state of mind can lead to mean that one can harm another to prove being fit or stronger. And to do as one pleases. This evolution theory has caused the unfortunate death of millions.

Evolution is built mainly on hypothesis, man's imaginations and drawings speculating how things arrived here on earth. Just a note, but evidence is not proof, as evidence is helpful in forming conclusions while proof concludes the matter altogether.

I can understand that perhaps the authors, book publishers and school boards may not have all the facts as well. But, there still should be freedom of choice, the right to choose from school teachings, what a person wants to learn and believe.

But once you have been told something for such a long time, you believe. Evolution is based on guesses to how all arrived. Perhaps some people didn't or do not still want to believe or accept that some supernatural being such as God created the heavens and the earth didn't mean they should build on this theory of evolution without any scientific evidence.

Unfortunately what we are taught through the public schools and the media, have convinced us that dinosaurs have been extinct for over 60 million years. Long before man set foot on earth. We are unable to imagine people and dinosaurs living together at the same time.

Dinosaurs were created no more than one day before mankind (sea) and same day as mankind (land), not millions of years earlier. And there is evidence to support that statement in creation science.

Since Charles Darwin popularized the theory of evolution in 1859, there has been no provable scientific evidence to backup his theory. The evolution theory has been developed through people writing books since Darwin's time. These books contain artist's drawings that show the names and the ages of layers on the earth as well as showing plants and animals arranged on a "family tree".

These pictures are not evidence, as anyone can draw a picture or construct a table and propose a theory, but that doesn't prove the theory. Even printing them for more than a hundred years doesn't prove anything. There isn't any first generation evidence on evolution or testable facts. Everything is based on assumption.

The evolution/creation controversy is not just a scientific issue. Scientific evidence strongly favors creation, but the information is censored from the public and certainly out of the public education. It's unfortunate that those who want to push evolution, can become very convincing with the words they use, and to the point where it convinces you, the individual how to think, what to think, and to not explore any other explanation for how you came about other than evolution.

A question I have is, if there is so much weight to the theory of evolution, why do evolutionists insist on attacking creation science? Just because you cannot see, touch, or hear the Creator, doesn't mean he doesn't exist. Just like air, because you can't see it, doesn't mean it isn't there.

There wouldn't have to be any debate if evolution was as clear-cut as they'd want us to believe. There isn't any real evidence for evolution as a scientific fact. People stick to theory as fact, even with no supporting evidence. Because theory has been presented as fact for so long, it is inconceivable for people to even consider the possibility that evolution is actually unsupportable.

Evolutionists have a difficult time answering several questions. Such as, why do fossils appear instantly in the fossil record? Why are there not many thousands of transitional fossils? There should be a very huge amount of transitional fossils, especially if it has been evolving for millions of years. With no concrete links of plants to animals, fish to amphibians, amphibians to reptiles, etc., the fossil record is composed mostly of large gaps. While Darwin predicted that the fossil record would show numerous transitional fossils, even a century and a half later, all we have are a handful of disputable examples.

Charles Darwin said, "*Why, if species have descended from other species by insensibly fine graduations, do we not everywhere see innumerable transitional forms? Why is not all nature in confusion instead of the species being, as we see them, well defined?*" Darwin also admitted, "*By this theory innumerable transitional forms must have existed, why do we not find them embedded in the crust of the earth?*" In answer to this, there aren't any fossils that exist showing any descending from one to another, even after 150 years.

Later in life Darwin wrote, "*I was a young man with uninformed ideas. I threw out queries, suggestions, wondering all the time over everything; and to my astonishment the ideas took like wildfire. People made a religion of them.*" And this is very apparent once 'natural selection' became the idea to how man; animal and plant came to be.

Marilyn Oakley

**Evidence for evolution**; (used with permission from www.allaboutcreation.org/) "Published by AllAboutGOD.com Ministries, M. Houdmann, P. Matthews-Rose, R. Niles, editors, 2002-07. Used by permission."

Quote **The Fossil Record:** (ex) Supposed "missing links" between distinct kinds of animals which can be extrapolated as transitions between kinds. For example, Archaeopteryx is thought to be a transition between reptiles and birds. (crit) There are no unambiguous transitional fossils. Archaeopteryx was thought to be a transition between reptile and bird because of its teeth and the claws on its wings. The fact is some fossil birds had teeth, and some didn't. Some reptiles have teeth, and some don't. Some mammals have teeth, and some don't. As far as claws on its wings, there are birds living today that have claws on their wings. Nevertheless, they are obviously birds, and no one disputes this. Besides, superficial similarities do not imply genetic relationship. "There is not one such fossil for which one could make a watertight argument. The reason is that statements about ancestry and descent are not applicable in the fossil record:" - Colin Patterson, Senior Paleontologist at the British Museum of Natural History and editor of a prestigious scientific journal. Patterson is a well-known expert having an intimate knowledge of the fossil record. (Reference: Colin Patterson, personal communication. Luther Sunderland, "Darwin's Enigma," 1988, p. 89.)

**Embryology:**
(ex) Embryos of different vertebrates look alike in their early stages, giving the superficial appearance of relationship. (crit) Embryos of different vertebrates DO NOT look alike in their early stages. "This idea was fathered by Ernest Haeckel, a German biologist who was so convinced that he had solved the riddle of life's unfolding that he doctored and faked his drawings of embryonic stages to prove his point." (William R. Fix, "The Bone Peddlers: Selling Evolution," 1984, p. 285.) Haeckel was exposed as a fraud in 1874 by Professor Wilhelm His. Nevertheless, Haeckel's fraudulent drawings (or similar representations) remain in high school and college biology textbooks to this day as evidence for evolution. Unquote

The Peppered Moth experiment was a hoax. Photographs of these peppered moths naturally camouflaged while at rest upon tree trunks were presented as evidence for natural selection. Biologists have known since the 1980's that peppered moths don't normally rest on tree trunks. The photographs found in the textbooks were staged using dead moths glued to tree trunks.

For myself to believe in evolution I'd like to see real answers to fundamental questions. Such as, can chemistry alone account for the origin of life on earth? Could undirected natural processes alone assemble the intricate structures found within living cells? What is the origin of the genetic information encoded in living organisms? Until evolutionists can come up with real answers to these fundamental questions, evolution should perhaps be kept in philosophy textbooks and taken out of biology textbooks where, unfortunately, it has become entrenched.

Creation science through overwhelming evidence points to the absurdity and the extreme impossibility of evolution. Geologists began to compile a geologic column, which is divided into a strata, as they had believed that evolution was a fact. But nowhere in the world does this geologic column actually exist. The problem with this geologic column, fossils had been found in the wrong strata, so the evolutionists try to explain it by calling it a Stratigraphic leak. With huge thicknesses of whole strata sometimes found in the wrong order, lead to even more problems. So the evolutionists call this *overthrusting* to try and explain it.

## I Was Blind But Now I See    Evolution - Creation

The recent findings that disprove this geologic column are:
1. In Arizona and Rhodesia, dinosaur pictographs have been found on cave and canyon walls drawn by man.
2. An ancient Mayan sculpture has been found of a bird that resembles the Archaeopteryx. There is a 130 million year problem here. If this geologic column is correct, the two should have never met.

Several trilobite fossils were found inside of the fossilized, footprint of a man who was wearing sandals on June 1, 1968 in Utah. But trilobites became extinct 230 million years ago according to the geologic column long before man came along.

The radioactive dating method is useless that scientists use for calculating the age of rocks and fossils. Carbon 14 is used to calculate the age of former living matters into 1000's of years and uranium lead used to calculate the age of earth into the millions of years. The problem with these methods; heating and deforming of rocks, percolation of water, exposure to neutrino, neutron, or cosmic radiation will alter the rates of decay, so making these methods totally useless. Even Potassium argon is a testing method.

Examples of these methods have shown that carbon 14 dated a living snail to being 2300 years old. Mind you, this snail is living, not dead or even fossilized. This same carbon 14 testing showed wood taken from a growing tree, not dead, nor fossilized as being 10,000 years old. The potassium-argon method found the Hawaiian lava flows to being 3 billion years old, when the lava was known to be less than 200 years old.

It is often pointed out that chemists have failed in their attempts to duplicate the spontaneous origin of life in the laboratory. Due to the ½ life of carbon 14, no objects over 50,000 years old should test positive for carbon 14. If a sample does test positive, it is great evidence that it is not millions of years old. If coal was formed over millions of years ago, it should not test positive for carbon 14. But that isn't what happens, there hasn't been any coal found completely empty of carbon 14.

Here is what I've found on Proofs of a young earth as compiled in Dr. David W. Cash's Manual on Creation of which I derived the following from and other sources.

The strength of the earth's magnetic field has been measured for over a 100 years. In a recent study, Dr. Thomas G. Barnes has shown that the earth's magnetic field is decaying at rate of 1/2 life every 1400 years; that means that the earth's magnetic field was twice as strong 1400 years ago as it is now. If the earth was even as old as 10, 000 years it's magnetic field would be as strong as a star! That is impossible! The earth could not possibly be any older than 10, 000 years! (Electric currents in the earth's core cause the earth's magnetic field. If we go back as far as 20, 000 years, we find that the heat produced by those currents would have melted the earth).

Meteoritic/Cosmic dust particles enter the earth's atmosphere at a rate of 14 million tons a year. If the earth were 5 billion years old there would be a layer of dust 182 feet thick over all the earth.

Evolutionists say that petroleum and natural gas took millions of years to develop. Recent experiments have shown that plant material has been converted into petroleum in only 20 minutes under the right temperature and pressure conditions. Also that petroleum and natural gas are contained at high pressures in underground reservoirs. Calculations based on measuring the cap rock has shown that this pressure could not be maintained longer than 10,000 years.

Evolutionists believe that coal was formed millions of years before man evolved (it is made of the metamorphosed remains of vegetation) they believe that these came from stagnant swamps.

Creationists believe that coal was formed due to the plants that were on the earth during the great flood. The types of plants involved and the texture of the coal shows that it was caused by turbulent waters, not stagnant swamps.

## Marilyn Oakley

In 1862, Macoupin County, Illinois, human male bones were discovered in a slate covered coal bed 90 feet underground. The bones were crusted with a carbonaceous deposit, which was easily scraped away to reveal white bone underneath. A similar skeleton found in a coal bed in Leicestershire, England, was reported in 1829. But people didn't exist when the coal was being formed, so you will not learn of these anomalies in school or encyclopedia, yet.

Due to the gravitational drag of the sun, moon and other factors, the rotation of the earth is gradually slowing. Its present rotation should be zero, if the earth is billions of years old, as the Evolutionist believes. Furthermore, the centrifugal force would have been so great that the continents would have been sent to the equator and the earth would have been as flat as a pancake if the earth were that old.

Even population proves a young earth. Let's take a look at that. Let's see, first the evolutionists believe that man has been on the earth for at least a million years. Creationists believe that man has been on the earth for only a few thousand years. Dr. Henry Morris calculated that an average growth of only ½% per year, which is only ¼ the present rate, would yield the population that we have today in only 4000 years. This is even taking into account war, disease, etc.

So, consider this, if the evolutionist is right and if the population has increased only ½% per year for a million years our population would be 10 to $2100^{th}$ power. Let's get an idea of about how many people that would be. Think about this; only 10 to the $130^{th}$ power electrons can be packed into the entire universe! Man has only been here a few 1000 years. We would for sure be looking for elbowroom if the evolutionist is right. End information from Dr. David W. Cash and other sources.

~~~

The Big Bang

The universe with all that it contains; space, time, matter and energy, the evolutionists would have us believe that it exploded from nothing. This is quite contrary to the First Law of Thermodynamics. The First Law of Thermodynamics is known as the law of energy conservation. It states that the energy can be converted from one form to another, but it can neither be created nor destroyed. In other words, it teaches that the universe did not create itself. If the evolutionists want to believe that the universe created itself, then where did the energy come from? So how can it be that no more energy is being created? (Genesis 2:3 explains that God ceased creating from that day forward.) How can it be that energy cannot be destroyed? (Hebrews 1:3 tells us that God is upholding all things by the word of his power.) So where did space, time, energy and matter come from in the first place? This ultimate question of origins remains unsolved by the evolutionists. This original explosion of everything from nothing, which in fact, complicates the evolutionary position, is not able to explain all of the complexity and fine-tuning in the universe. Evolutionists have formed their conclusions and are now looking for the missing data.

Evolutionists contend that complexity has developed from simplicity over time. They view time as their solution. Hard science however, tells us that time is the enemy of complexity. This fact has been so well documented that is has obtained the stature of a physical law, the Second Law of Thermodynamics. The second law states that every system that is left to its own devices tends to move from order to disorder. So, it can be said, that the universe is proceeding in a downward degenerating direction of decreasing organization. That leads to eventually the universe will die a heat death and it will waste away to nothing. But evolution teaches a process that continues upward, constantly becoming more orderly and improved. This is impossible.

Let's look at Dr. David W. Cash's example of how things cannot continue in an upward more orderly and improved process; Quote – "For instance, if you leave out a pile of building material such as lumber, brick, etc. in a field where energy from the sun can reach it and you come back a few million or billion years later, would you find a building there? Hardly! What you would find would be nothing as it all would have deteriorated and gone back to dust. The Evolutionists believe just the opposite. They feel that it you leave all the components of a creature in an area where energy can reach it under the right conditions, it will become a living creature. What do you think?

The mere fact that we get old, our body breaks down and we finally die is living proof of the second law of thermodynamics. British astronomer Arthur Eddington quotes: "*If your theory is found to be against the second law of thermodynamics I can give you no hope; there is nothing for it but to collapse in deepest humiliation*". The truth of the matter is that the second law of thermodynamics was ordered by God in Gen 3: 17 - 19 and can only be changed by God (who will one day do so in Romans 8: 18 - 23)." Unquote

Big Bang Hypothesis (1948) Astronomers were totally buffaloed as to where matter and stars came from. In desperation, *George Gamow and two associates dreamed up the astonishing concept that *an explosion of nothing produced hydrogen and helium, which then shot outward, then turned and began circling and pushing itself into our present highly organized stars and galactic systems*. This far-fetched theory has repeatedly been opposed by a number of scientists (*G. Burbidge, "Was There Really a Big Bang?" in Nature 233, 1971, pp. 36, 39)*. By the 1980s, astronomers which continued to oppose the theory began to be relieved of their research time at major observatories ("*Companion Galaxies Match Quasar Redshifts: The Debate Goes On*," Physics Today, 37:17, December 1984). In spite of clear evidence that the theory is unscientific and unworkable, evolutionists refuse to abandon it.

Read about **Catastrophism** Used with Permission from ttp://www.allaboutcreation.org/ "Published by AllAboutGOD.com Ministries, M. Houdmann, P. Matthews-Rose, R. Niles, editors, 2002-07. Used by permission."

Quote Catastrophism is contrary to Uniformitarianism, the accepted geological doctrine for over 150 years. Uniformitarianism states that current geologic processes, occurring at the same rates observed today, in the same manner, account for all of earth's geological features. As present processes are thought to explain all past events, the Uniformitarianism slogan is "the present is the key to the past." Uniformitarianism ignores the possibility of past cataclysmic activity upon the surface of the earth. James Hutton first purposed the doctrine of uniformity in his publication, *Theory of the Earth* (1785). Sir Charles Lyell endorsed Uniformitarianism in his work, *Principles of Geology* (1830). Uniformitarianism is fundamental to Lyell's geologic column. Uniformitarianism and the geologic column, both of which assume uniformity, have been disputed in recent years by geologic features such as poly-strata fossils, misplaced fossils, missing layers and misplaced layers (including layers in reverse order or "ancient" layers found above "modern" layers). Uniformitarianism, together with the Geologic Column presupposed by Lyell based on uniformity, have been disproved by those geologic features stated above.

Furthermore, observed cataclysmic events such as the eruption of Mt. St. Helens in 1980 have leant credibility to Catastrophism. Prior to the introduction of Uniformitarianism, Catastrophism was the accepted geological doctrine. Once again, Catastrophism is becoming accepted as an accurate interpretation of earth's geologic history.

In regards to Uniformitarianism, Warren D. Allmon writes, "As is now increasingly acknowledged, however, Lyell also sold geology some snake oil. He convinced geologists that because physical laws are constant in time and space and current processes should be consulted before resorting to unseen processes, it necessarily follows that all past processes acted at essentially their current rates (that is, those observed in historical time). This extreme gradualism has led to numerous unfortunate consequences, including the rejection of sudden or catastrophic events in the face of positive evidence for them, for no reason other than that they were not gradual." ("Post Gradualism", *Science*, vol. 262, October 1, 1993, pg. 122).

~~~

Is the Bible true? By inspiration of God. The Bible claims to be given by inspiration of God, but it is unique in that it offers substantial evidence to back its claims. The Test of Prophecy, unquestionably, the single greatest evidence lending to the veracity of the Bible's claims of divine inspiration is the fulfillment of Bible prophecy. The vast majorities of these prophesies have already come to pass and can be verified by secular history. Archaeology, not only does the Bible miraculously foretell the future, it also recounts the distant past with great accuracy.

Dr. John McRay, Professor of New Testament and Archaeology at Wheaton University in Illinois, explains, "*The general consensus of both liberal and conservative scholars is that Luke is very accurate as a historian. He's erudite, he's eloquent, his Greek approaches classical quality, he writes as an educated man, and archaeological discoveries are showing over and over again that Luke is accurate in what he has to say.*" (John McRay, quoted by Lee Strobel, *The Case For Christ*, Zondervan, 1998, p. 129.) Luke who authored approximately one-quarter of the entire New Testament is regarded as an authoritative historian -- one of the greatest of antiquity. And Sir William Ramsey, one of the greatest archaeologists of modern times, declared, "*Luke is a historian of the first rank.*" (Sir William Ramsey, *The Bearing of Recent Discovery on the Trustworthiness of the New Testament*, 1915, p. 222. Unquote

Macroevolution is called the vertical transformation of one kind of organism to another. Just to let you know, this cannot and does not occur. Dogs do not become horses and parakeets do not become crows. Even certain species of animals have their own boundaries. Let me give you some examples: A horse plus a donkey equals a sterile mule, a lion plus a tiger equals a sterile tiger. So, given those examples, even animals cannot reproduce a new race of animal. If evolution played a role in our being today, where are the links between non-living matter to protozoans? From protozoans to metazoan invertebrates? From metazoan invertebrates to amphibians? From amphibians to reptiles? From reptiles to birds? From birds to fur-bearing quadrupeds? From quadrupeds to apes? From apes to man? They don't exist!

That's the big problem and again repeating what Darwin confesses: "*As by this theory, innumerable transitional forms must have existed. Why do we not find them imbedded in the crust of the earth?*" Even many kindergarten students can answer that question: "*In the beginning God created the heaven and the earth*".

This problem caused a panic among the evolutionists, so they had to come up with another theory to cover for this mistake. Now they call the new theory "punctuated equilibrium". Which means, they are trying to say now that evolution occurred in big leaps instead of tiny ones. So, can I safely conclude that basically, they are saying is that at one time a reptile laid an egg and a bird hatched from it? So, on one hand, in the beginning according to the evolutionist, we slowly evolved. But, they couldn't come up with any evidence and no proof. So, now they are going to say all living creatures including man now evolved in big leaps! And still they say this with no scientific evidence to support it. Evolution doesn't leave much for myself to believe it and in my opinion should be in the science fiction section.

If fossils had been formed on a large scale as they say happened millions of years ago, how come this process doesn't continue? Why aren't any fossils being formed today? If evolution is the answer, then evolution should still be in process. It doesn't simply stop, just because. The evolutionist is trying to convince us that we evolved, but then is he going to try and convince us that it suddenly stopped? Since when and how come it stopped? Why did it stop and where did it stop? Who said it stopped and are we so perfect, we don't need to evolve anymore? If evolution is the answer, then the human of today, should have evolved from the humans of a 1000 years ago, of 2000 years ago. So, if evolution is the theory, what changes are there between man of today and man of long ago, once supposedly we stopped evolving after we left the form of an ape?

And this is what Dr. Carl Sagan once proclaimed; Quote "*We are the product of 4.5 billion years of fortuitous, slow biological evolution. There is no reason to think that the evolutionary process has stopped. Man is a transitional animal. He is not the climax of creation.*" Unquote

It makes me wonder how much more we can evolve? Does this mean my child is still evolving over me? Will their children still evolve a bit farther? Is there like a certain gene (DNA) that will go into my child from my body, and he will carry it in his until he has children, and then pass it along to them... so that eventually this gene will somewhere along the line mutate or evolve into something else? So in reference to this gentleman's proclamation, how then are we today different than man a 1000 or 2000 years ago? That's if we are still slowly evolving as each year, decade, and century passes. But, then again, I had read that the process of evolution stopped 100,000 years ago. Isn't evolution contradicting themselves?

The Fossil Record

There are two periods, the Precambrian period and the Cambrian period. In the Precambrian period, there is total absence of multicellular life forms (fossils) found in the lower 2/3 of the earth's crust. If we had evolved from tiny, simple life forms, you would think that there would be an abundance of these mutilcellular forms (fossils) in the strata of the earth. How about the Cambrian period? Well, advanced life does appear in abundance during this period. It contains the oldest rocks in which complex fossils are found. These rock fossils contain many millions of highly advanced and well-developed life forms. The record indicates that life appeared suddenly in tremendous complexity, great diversity and unbelievable abundance without evolving from any ancestors. While in the Precambrian period, the fossil record only has sparse, unicellular fossils. There doesn't seem to be a connection between unicellular organisms and all other life form.

As for the fossil record, there is no gradual evolution at all. If everything did develop through evolution, there would be some basic creatures that would have existed in familiarity of the creatures we know today. But there isn't any evidence of any creature turning (evolving) in the fossil record at all.

Even Charles Darwin admitted this: "*Not one change of species into another is on record... We cannot prove that a single species has been changed.*"

But on the contrary, the fossil record does strongly support the Biblical principle of reproduction "after its kind". Let's look at Genesis 1 11-12, 22-25. Note that it constantly refers to "after its kind" when it mentions each type of animal. Let me give an example, there are over 200 varieties of dogs, but they are all dogs. How about the 9 foot giant Anakin of ancient Palestine found in Numbers 13: 28-33 and Deuteronomy 2: 10, 11, 21 in comparison to the 4 foot pygmies of central Africa? Although there may be a 5-foot difference in height, they are all men.

Darwin was at least honest enough to acknowledge the unavoidable problem of all the missing links. There has been no real discovery of any credible transitional fossils. This record continues to be composed of mainly gaps.

The Flood was a real historical event and the earth's crust bears witness to this in many compelling ways. Billions of dead things buried in sedimentation ("laid-down-by-water rock") found all over the earth. Geologist Dr. John Morris explains, "*Sedimentary rocks, by definition, are laid down as sediments by moving fluids, are made up of pieces of* rock or other material which existed somewhere else, and were eroded or dissolved and redeposited in their present location." Over 70% of the earth's surface rock is sedimentary rock (the rest of earth's surface rock is volcanic igneous and metamorphic rock).

Let's look more into fossil graveyards. Did you know that in the Cumberland Cavern in Maryland, it contains the remains of animals from at least 5 different regions of the world? How do those evolutionists explain this? In fact, there are caves and caverns all over the world that are packed with masses of fossils. Even being thrown together in disorderly masses, these various animals had come from widely separated regions of the world. The only explanation can be in terms of a worldwide flood.

If fossils were to continue, then the fish that dies would become a fossil, but instead, it sinks to the bottom, decays and gets eaten by other fish. No chance to fossilize at all. Let's look at buffalos; some herds were big enough to cover an entire state, but show there is no trace of the millions that once existed, and were slaughtered all over the plains just a couple of generations ago. Why is that? Because fossilization did not occur. Fossilization occurred as a result of the great flood of Genesis. This occurred about 2300 B.C. and is recorded in Genesis 7:23.

Polystrare fossil trees, often more than 20 feet have been extending through several rock layers. If you listen to the evolutionists, who says this strata was formed over millions of years, then the tree would be decomposed while it waits for the strata to form. But these fossil trees didn't decompose, making obvious the correct explanation that the strata was formed instantly and the tree was deposited in the strata at the same time it was formed. For something to become fossilized, it has to be buried either causing its death or immediately after. Otherwise, it would decay or be scavenged by animals and there would be nothing to fossilize. These type fossils are some of the best evidence rock layers haven't been laid down over many millions of years.

What did William Jennings Bryan have to say? He was a presidential candidate for the Democratic Party. Jennings was an honored statesman and sought after speaker, yet it was his rally against Darwinism in the Scopes Monkey Trials he is most remembered for. He once said, "*Evolution seems to close the heart to some of the plainest spiritual truths while it opens the mind to the wildest guesses advanced in the name of science.*"

How about these statements? - "*I could prove God statistically; take the human body alone; the chance that all the functions of the individual would just happen, is a statistical monstrosity.*" George Gallup, Famous statistician.

"*It is as impossible that pure incogitative matter should produce a thinking intelligent Being, as that nothing should of itself produce Matter.*" John Locke, Essay Concerning Human Understanding, 1690, IV, x. 10.

Also, fossils (human/ape) are usually fragmented and incomplete. Any conjecture based on them is likely to be completely speculative. Reconstruction of these fragments or incomplete fossils is prepared speculatively precisely to validate the evolutionary thesis. Since these fossils are not complete, man can reconstruct them to look the way he wants. The half-man half-ape reconstructions do not exist as whole fossils or even fossil traces. But they are very convincing when shown as drawings, in magazines and textbooks. Theories have, in the past, clearly reflected our current ideologies instead of the actual data.

Reconstructions based on bone remains can only reveal the most general characteristics of the creature, since the really distinctive morphological features of any animal are soft tissues, which quickly vanish after death. Therefore, due to the speculative nature of the interpretation of the soft tissues, the reconstructed drawings or models become totally dependent on the imagination of the person producing them. Earnst A. Hooten from Harvard University explains the situation like this:

I Was Blind But Now I See   Evolution - Creation

To attempt to restore the soft parts is an even more hazardous undertaking. The lips, the eyes, the ears, and the nasal tip leave no clues on the underlying bony parts. You can with equal facility model on a Neanderthaloid skull the features of a chimpanzee or the lineaments of a philosopher. These alleged restorations of ancient types of man have very little if any scientific value and are likely only to mislead the public... So put not your trust in reconstructions.

Three different reconstructions based on the same skull.

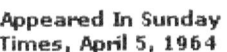

Appeared In Sunday Times, April 5, 1964

Maurica Wilson's drawing

Parker's reconstruction N.Geographic, September 1960

As a matter of fact, evolutionists invent such "preposterous stories" that they even ascribe different faces to the same skull. For example, the three different reconstructed drawings made for the fossil named Australopithecus robustus (Zinjanthropus), are a famous example of such forgery.

There aren't any mechanisms in nature to lead the living beings to evolve and that living species came into existence not as the result of an evolutionary process, but rather emerged all of a sudden in their present perfect structure. That is, they were created individually. Therefore, it is obvious that "human evolution", too, is a story that has never taken place.

What, then, do the evolutionists propose as the basis for this story? This basis is the existence of plenty of fossils on which the evolutionists are able to build up imaginary interpretations. Throughout history, more than 6,000 ape species have lived and most of them have become extinct. Today, only 120 ape species live on the earth. These approximately 6,000 ape species, most of which are extinct, constitute a rich resource for the evolutionists.

The evolutionists wrote the scenario of human evolution by arranging some of the skulls that suited their purpose in an order from the smallest to the biggest and scattering the skulls of some extinct human races among them. According to this scenario, men and modern apes have common ancestors. These creatures evolved in time and some of them became the apes of today while another group that followed another branch of evolution became the men of today.

However, all the paleontological, anatomical and biological findings have demonstrated that this claim of evolution is as fictitious and invalid as all the others. No sound or real evidence has been put forward to prove that there is a relationship between man and ape, except forgeries, distortions, and misleading drawings and comments.

The fossil record indicates to us that throughout history, men have been men and apes have been apes. Some of the fossils the evolutionists claim to be the ancestors of man, belong to human races that lived until very recently-about 10,000 years ago-and then disappeared. Moreover, many human communities currently living have the same physical appearance and characteristics as these extinct human races, which the evolutionists claim to be the ancestors of men. All these are clear proof that man has never gone through an evolutionary process at any period in history.

The most important of all is that there are numerous anatomical differences between apes and men and none of them are of the kind to come into existence through an evolutionary process. "Bipedality" is one of them. Bipedality is peculiar to man and it is one of the most important traits that distinguishes man from other animals.

The first Ramapithecus fossil found, which relied on only two pieces of jawbone, drew the imaginary evolutionists conclusion to his family and the environment they lived in.

The complexity of design

The human eye is furnished with automatic aiming, automatic focusing, automatic aperture adjustment. It can function in almost total darkness and in bright sunlight as well. It can see an object as fine as a diameter of a hair and make about 100,000 motions a day. Then, while we sleep it performs its own maintenance work. Scientists still do not fully understand how it works. It doesn't seem to likely that the human eye could have evolved into such an intricate organ.

As Dr. David W. Cash said in his Manual On Creation: Quote "There are many animals that migrate from one place to another depending on the seasons and time of year often traveling for thousands of miles. They use no electronic equipment, no compass, radio, direction finders, etc. They cannot read a map and are certainly unable to stop and ask directions as we can. Yet they never miss their destination and always arrive on time. How do they do this? The answer to this has long baffled the Evolutionist, for according to his beliefs, this has to be a miracle. It is indeed a miracle, but not an unexplainable one. The creationist understands that this is all in God's plan. Only a supernatural being could create such an intricate navigation system." Unquote

How about the case of the white-throated warbler, whose young leave several weeks after the parents do. How do they find their way several 1000 miles away?

Dr. Cash included in his Manual On Creation; Quote 'Experiments have shown that within the brains of these birds is the inherited knowledge of how to tell latitude, longitude, and direction by the stars, plus a calendar, clock, and all the necessary navigational data needed to make such incredible journeys to the exact locations of their parents.' Unquote

How could this remarkable ability evolve a little at a time? Being able to navigate across only half of the ocean would not be too likely. In conclusion, these animals were created and designed with these abilities, and not just simply evolved.

What about visual beauty? Let's look at what Dr. Cash said - Quote; "Visual beauty must have a practical purpose according to Evolutionist or else it would have never evolved. However, there are practically countless cases of creatures that have visual beauty that serves absolutely no purpose at all except for the personal gratification of God and man.

For example: The abyssal fish (Rhodicthys) is of a bright red color. Yet it lives in total darkness 1 1/2 miles below the surface of the ocean. There is no reason for it (according to the Evolutionist) for it to have this color. It serves no practical purpose. Even the eggs of many deep-sea creatures are brilliantly colored.

How about this: Why are the insides of some shells brilliantly colored? Why are the insides of some baby bird's mouths brilliantly colored? Also if visual beauty is a sign of evolutionary progression, why is it that the lower forms of life display greater visual beauty than man?

Psalm 19: 1 "The heavens declare the glory of God; and the firmament showeth his handiwork." (That's why!!!)" Unquote

Evolutionists believe that man and apes evolved and creationists maintain that man was supernaturally created.

I found this as well in Dr. David W. Cash's Manual On Creation.

A Nebraska man was discovered in 1922 by Harold Cooke in the Pliocene deposits of Nebraska. Many books were written about this Nebraska man which supposedly lived over a million years ago. The evidence for Nebraska man was used in the infamous

I Was Blind But Now I See   Evolution - Creation

scopes evolution trial in Dayton, Tennessee in 1925. William Jennings Bryan was confronted with a group of experts who blew him away with "facts" of Nebraska man. Mr. Bryan was made a total mockery of and laughed at. But where did Nebraska man really come from? Would you believe a tooth? Yes, only a tooth. This was proof positive that a prehistoric race existed because of a single tooth! Years after the scopes trial, the entire skeleton that the tooth belonged was dug up and it was found to be that of an extinct pig!! Needless to say, very little publicity was given to it by the news media when this happened.

The Southwest Colorado man was a similar case like that of the Nebraska man. He was a so - called "missing link" reconstructed around nothing but a single tooth. It is now known that this tooth belonged to a horse! How about the Java ape-man who is one of the most famous of all the anthropoids. He was discovered in 1891 by Dr. Eugene duBois, a very fervent Evolutionist. This "ape man" was reconstructed from a small piece of the top of a skull, a fragment of a left thighbone, and three molar teeth. 24 scientists met to evaluate this find: 10 said it was a man, 7 said it was an ape, 7 more had no idea (they believed that it belonged to a no longer missing link)

But there were problems: The bones that were used were not even together, they were scattered some 70 feet apart. They were not even found at the same time; it took a year to find these fragments. To make matters worse, these bones were found in an old riverbed mixed in with the bones of extinct animals. So we ask the question of how can these scientists be certain that these bones all belonged to the same animal? If they were indeed 750, 000 years old and were not petrified, how did they last so long without disintegrating? How could you accurately reconstruct an entire skeleton with such tiny pieces of evidence?

Well, as it turns out, even Dr. DuBois, the finder himself, concluded that these were the bones of a gibbon. But it was already too late. The "reconstructed" skeleton was on display at a museum and college textbooks were already written singing the praises of this phony "ape man".

Another of these Java "ape men" was discovered in 1926. This was another darling of the scientific community like the first one. This one turned out to be the knee bone of an extinct elephant.

The remains of Piltdown man were supposedly found in 1912 by Charles Dawson, an amateur fossilogist. He produced some bones, teeth, and primitive tools, which he said he found in a gravel pit near Piltdown, Sussex, England. He took the remains to Dr. Arthur Smith Woodward, an well known paleontologist at the British museum. These remains created a flurry of activity among scientists who immediately dated the remains at 500, 000 years old. Literature flooded the bookshelves as this was hailed as the most remarkable of all finds. Over 500 doctoral dissertations were written on "Piltdown man". The great "missing link " was found!! Or was it???

In October of 1956, the worst nightmare of the Evolutionist was about to happen. Reader's Digest came out with an article summarized from Popular Science monthly called "The Great Piltdown Hoax". Using a new method to date bones based upon fluoride absorption, the Piltdown man was found to be a fraud!! Further study revealed that the jaw-bone used in Piltdown man actually belonged to an ape that had died only 50 years previously. The teeth were filed down, and both teeth and bones were discolored with dichromate of potash.

The man who was responsible for placing the bones in the gravel pit was a man named Teilhard de Chardin S. J. He had authored several books attempting to harmonize evolution with Christianity. Frustrated by the lack of evidence for Darwin's theory, he thought he would assist by inventing this "missing link".

Marilyn Oakley

How true Romans 1: 21 is when it says "professing themselves to be wise, they became fools."

The specimen Neanderthal man was discovered around the turn of the century in the Neanderthal valley of Dusseldorf, Germany. He was originally portrayed as a semi - erect, barrel chested, brute which was to be the final missing link between ape and man.

After many other Neanderthal skeletons were discovered, it was later found that Neanderthal man was fully human. In fact, he could even be considered superhuman due to the fact that his cranial capacity was 13% bigger than modern humans.

The reason that many thought he was a primitive missing link was due to the fact that the first specimen that was studied was crippled with osteo - arthritis and rickets giving him that semi erect, "cave man" appearance. Today Neanderthal man is classified as Homo - Sapiens, which means it is totally human.

Lucy was a specimen that comes from a group of fossils called "Australopithecines". Lucy was a female of this group constructed from a 40% complete skeleton. Lucy was discovered in the Afar area of Ethiopia during studies conducted from 1972 - 1977. National Geographic reported quoted Johanson as saying: "The angle of the thigh bone and the flattened surface at its knee joint end. . . . proved she walked on two legs. "

However the knee joint end of the femur was severely crushed. Therefore assuming that "Lucy" walked upright is merely guessing. Anatomist Charles Oxnard, using a computer technique for analysis of skeletal relationships, has concluded that "Lucy" and other Australopithecines did not walk upright, at least not as humans do, but as chimpanzees.

Therefore, there is no valid scientific evidence that "Lucy" or any of her species walked upright as humans do. They are most likely a variety of apes. There are fossils of humans that have been found to be older than "Lucy". Therefore, she cannot be one of our "evolutionary" ancestors. Fully aware of the knowledge of the total absurdity of evolution, most of our scientists of today still choose to accept its preposterous theories. Not only do they blindly follow after dead - end suppositions, but they have led the young minds of our world down those same paths. This is the greatest tragedy of all. Dr. George Wald, winner of the 1967 Nobel peace prize in science, wrote: "When it comes to the origin of life on this earth, there are only two possibilities: creation or spontaneous generation (evolution). There is no third way. Spontaneous generation was disproved 100 years ago, but that leads us only to one other conclusion: that of supernatural creation. We cannot accept that on philosophical grounds (personal reasons); therefore, we choose to believe the impossible: that life arose spontaneously by chance. "

"The fool hath said in his heart, there is no God." (Psalm 14: 1) Unquote Dr.Cash

Used with Permission http://www.evolution-facts.org/

Quote "Let's look at some actual evidence of early man on our planet. They were real human beings and where they were found disproves the evolution theory. **The Guadeloupe woman.** In 1812, on the Caribbean island of Guadeloupe, a fully human skeleton was found, lacking only the head and feet. It was found inside extremely hard, very old limestone, which was part of a formation over a mile in length. You will not find the Guadeloupe woman mentioned in the textbooks, since this find would disprove evolutionary strata dating. **The Caveras skull.** In 1876, 130 feet below ground, a skull was found in the "2 million-year-old" Pliocene strata. It was certified as completely mineralized, yet totally human. Dozens of stone mortars, bowls, and other man-made artifacts were found near this skull. **The Castinedolo skull.** A group of perfectly human ancient skulls were found in Castinedolo, Italy, and, with the Caveras Skull, are considered among the most ancient skulls. Yet they are perfectly human. **The Moab Skeletons.** Two skeletons were found in Cretaceous rock (supposedly dated at 100 million years) in Moab, Utah, about 15 feet below the surface. Both skeletons were definitely human and ancient. They had been undisturbed till they were found. When tested for age, they were only a few thousand years old.

Evolutionists theorize that man did not evolve until the late Tertiary Period, and cannot be over one to three million years old. But human footprints have been found in very old rock strata. These are human footprints, not ape prints. These prints disprove evolutionary theories about rock strata—and reveal it is quite young, and place dinosaurs as living at the same time when people did. The prints also reveal that giants once lived on our planet.

Laetoli tracks. At a site in Kenya, called Laetoli, 30 miles south of Olduvai Gorge, Mary Leakey discovered human footprints in 1977. Although some evolutionists reject them as human, other scientists recognize them to be clearly human—and therefore date those who made the tracks to be 3.75 million years ago. But evolutionists teach that no people lived back then.

At about the same time, Mary Leakey and Dr. Johanson found human teeth and jawbones from around the same 3.75 million-year period. **The Gediz track.** A footprint found in volcanic ash, near Demirkopru, Turkey, was found in 1970. The track of a running man was found in strata dated by evolutionists at 250,000 years in the past.

The Glen Rose tracks. A remarkable number of human tracks have been found in a Cretaceous limestone formation near Glen Rose, Texas. Many are of giant men. The prints have been found in the bed of Paluxy River, when it is dry in the summer. Some are next to, on top of, or under dinosaur tracks.

The Glen Rose tracks are 15 inches long [38.1 cm], and were probably made by people 8.3 feet [25,38 dm] tall. Some, 21½ inches [54.6 cm] long, would have been made by people 11.8 feet [25.38 dm] tall.

R.T. Bird, a paleontologist with the American Museum of Natural History, also found a trail of Brontosaurus tracks which were shipped to the museum. That means people were alive when the dinosaurs were! Some human tracks overlaid the dinosaur tracks, and some were found in layers below the dinosaurs.

The Paluxy Branch. In August 1978, Fred Beierie spent the afternoon searching for tracks in the Paluxy riverbed. He found a tree branch encased in Cretaceous stone, with only the upper part showing. So it was as old as the tracks.

Beierie sent a sample of the wood to *Reisner Berg of UCLA, who tested it by radiodating at 12,800 years. Corrected, it would yield a date agreeing with the Flood. (Carbon 14 dating tends to skew toward greater age on older dates, because of atmospheric differences back then. See Dating of Time in Evolution http://www.pathlights.com/ce_encyclopedia/Encyclopedia/06dat1.htm for details.)

That would date both the giants and the dinosaurs as being recent.

The Alamogordo tracks. Near Alamogordo, New Mexico, 13 giant tracks, each about 22 inches [55.8 cm] long were found. The stride is from four to five feet [121.9-152.4 cm].
The Arizona tracks. Tracks of a barefoot human child were found, in the late 1960s, alongside some dinosaur tracks. The location was the Moenkopi Wash, near the little Colorado River in northern Arizona.

In 1984, similar tracks were found not far from the Moenkopi site. Many human tracks, dinosaur tracks, and a handprint of a child that had fallen. More adult tracks were found in 1986. The Arizona tracks are located in the Glen Canyon geological Group, which is part of late Triassic to early Jurassic strata and supposedly 175 to 100 million years old. In addition to 300 tridactyle dinosaur tracks, sheep tracks, bivalve prints, large amphibian and lungfish marks have been found. Over 60 human tracks have been mapped and photographed.

Marilyn Oakley

Human remains and man-made objects have been found in coal and rock—where they should not be found. The evidence disproves evolutionary theories about the age of rock strata. As far back as we can trace, people were people. They were not apes.

Man-made objects in rock. An iron nail was found in a Cretaceous block from the Mesozoic era (mid-1800s). A gold thread was found in stone in England (1844). An iron nail was found in quartz in California (1851). A silver vessel was found in solid rock in Massachusetts (1851). The mold of a metal screw was found in a chunk of feldspar (1851). An intricately carved and inlaid metal bowl was found in solid rock (1852). An iron nail was found in rock in a Peruvian mine by Spanish conquistadores (1572).

Man-made objects found in the ground. A doll was found near Nampa, Idaho (1889). A bronze coin was found 114 feet below the surface near Chillicothe, Illinois (1871). This means there were coins in ancient times in America! A paving tile was found in a "25 million-year-old" Miocene formation in Plauteau City, Colorado (1936).

Several discoveries were made during the California gold rush (1849-1850s). A prehistoric mining shaft, 210 feet [640 dm] below the surface in solid rock was found. A mortar for grinding gold ore was found at a depth of 300 feet [914 dm] in a mining tunnel. A human skull was also found at a depth of 130 feet [396 dm] under five beds of lava and tufa. Bones of camels, rhinoceroses, hippopotamuses, horses, and other animals were also found in California. The findings are almost always in gold-bearing rock or gravel.

Man-made markings on petrified wood. Evolutionists declare that petrified wood is millions of year old, yet humans have worked with it. Hand-worked petrified wood was found in India. It was shaped prior to fossilization. Prior to mineralization, several petrified pieces of wood had been hacked with a cutting tool. The wood was dated to the Pliocene Epoch, before humans were supposed to have lived.

Man-made markings on bones. At a site near Paris, France, fossilized rhinoceros bones had human cutting marks on them. No rhinos have been in Europe in recorded history. Another rhino bone, cut by a sharp tool, was found in Ireland. Two saurian bones were found in a Jurassic deposit.

Ancient cultures. Mankind appears to have migrated from a central point, located somewhere in the Near East or Asia Minor. This would agree with the conditions following the Flood and the fact that the Ark came to rest in eastern Turkey (see Genesis 8-9). In the Near East we find the earliest pottery, domestication of plants and animals, metalworking, towns and cities, and the earliest writing. The earliest authentic dates only go back to about 5,000 years ago. If man developed a "modern brain" 500,000 to 100,000 years ago (as the evolutionists tell us), then why did mankind wait till 5,000 years ago to begin using it?

Evolutionists say the first man came from central Africa, yet all the earliest human cultural activities began in the Near East." Unquote

As I said, there was a primitive understanding of science in Darwin's time. That understanding was firmly fixed on assuming that very simple structures made living things. The theory that non-living matter could bring about living forms became known as spontaneous generation, and that theory had been widely accepted.

Leftover food was thought to create insects and that wheat had produced living mice. They even did experiments on this by placing wheat on a dirty piece of cloth. They watched to see if mice would emerge alive from this non-living dirty piece of cloth.

Maggots that appeared from meat were thought to be spontaneous generation. But they realized later, that wasn't the case. Maggots came about as a result of flies laying their larvae on the meat.

In the era when Darwin's *Origin Of Species* was written, it became widespread that bacteria could come from inanimate matter.

I Was Blind But Now I See   Evolution - Creation

Louis Pasteur had made an announcement 5 years after Darwin's book publication. After long studies and experiments, he announced that his results disproved spontaneous generation, which was the foundation of Darwin's theory. At the Sorbonne in 1864, Louis Pasteur, in his triumphal lecture said, *"Never will the doctrine of spontaneous generation recover from the mortal blow struck by this simple experiment."* For a long time, advocates of evolution refused to accept Pasteur's findings.

How about the world of cells? A cell is a very complex factory. There are various processes within the cell. The processes are all part of one complex system. One process cannot function without the other processes. This means that they are all or nothing, because either everything is there and it works, or something is missing and it cannot work.

Michael Behe, biophysics professor at Lehigh University in Pennsylvania, reasons that a cell can only function as a complete unit. That means a cell is unable to function if it gradually comes into existence by means of evolution. He uses a mousetrap as an example. A mousetrap does not function if parts are missing. All the various parts have to be assembled in the right order for the mousetrap to work smoothly. In this sense, a cell can only function if all the 'parts' (or systems) are present within the cell. That's what we call the 'irreducible complexity' of a cell. If only one minor fragment is missing, the cell is unable to function.

~~~

From ObviousTruths.com – Used with Permission. Quote "So the question is: How could a cell evolve from a 'simple' to a 'complex' form if it only functions as a whole, all systems included?

Now, this is the cell under present consideration. If the cell is a complex "factory" where all parts must be in place in order to function, how can any "honest" person think that the hundreds of systems that make up the human body "gradually" came together over millions of years? How could the digestive system, lungs, eyes, ears, and the brain have lived independently of the other systems while they "evolved" since they must all work together as a unit to function?

Evolutionists will say that all of the complex organisms we see on the face of the earth occurred by random chance with no intelligence to guide or create these organisms. However, how did the brain develop by random chance with no intelligent designer?" Unquote

So, is someone going to tell me that the complexity of the eye, the lungs, our hearing, the heart, the brain and the nervous system are part of an evolution theory? That these parts of the body took millions of years to develop to the point they are at today? So, ah… natural selection, time, chance and a wee bit o'luck are responsible for my body as a whole, where everything works, finally, after millions and millions of years?

The popular media often portrays the creation vs. evolution debate as science vs. religion, with creation being religious and evolution being scientific. In an ironic twist, it's the creationists who have a solid empirical basis for their theory, while the evolutionists are left clinging to their convictions by faith.

Here are a few Famous Evolutionists quotes: Used with Permission
(You may read more quotes here: http://www.anointed-one.net/quotes.html)

"Undeniably, the fossil record has provided disappointingly few gradual series. The origins of many groups are still not documented at all." (Futuyma, D., Science on Trial: The Case for Evolution, 1983, p. 190-191)

Marilyn Oakley

"The main problem with such phyletic gradualism is that the fossil record provides so little evidence for it. Very rarely can we trace the gradual transformation of one entire species into another through a finely graded sequence of intermediary forms." (Gould, S.J. Luria, S.E. & Singer, S., A View of Life, 1981, p. 641)

"Few paleontologists have, I think ever supposed that fossils, by themselves, provide grounds for the conclusion that evolution has occurred. An examination of the work of those paleontologists who have been particularly concerned with the relationship between paleontology and evolutionary theory, for example that of G. G. Simpson and S. J. Gould, reveals a mindfulness of the fact that the record of evolution, like any other historical record, must be construed within a complex of particular and general preconceptions not the least of which is the hypothesis that evolution has occurred. ...The fossil record doesn't even provide any evidence in support of Darwinian theory except in the weak sense that the fossil record is compatible with it, just as it is compatible with other evolutionary theories, and revolutionary theories and special creationist theories and even historical theories." (Kitts, David B., "Search for the Holy Transformation," review of Evolution of Living Organisms, by Pierre-P. Grassé, Paleobiology, vol. 5, 1979, p. 353-354)

"In spite of these examples, it remains true, as every paleontologist knows, that most new species, genera, and families, and that nearly all new categories above the level of families, appear in the record suddenly and are not led up to by known, gradual, completely continuous transitional sequences." (Simpson, George Gaylord, The Major Features of Evolution, 1953, p. 360)

"The record certainly did not reveal gradual transformations of structure in the course of time. On the contrary, it showed that species generally remained constant throughout their history and were replaced quite suddenly by significantly different forms. New types or classes seemed to appear fully formed, with no sign of an evolutionary trend by which they could have emerged from an earlier type." (Bowler, Evolution: The History of an Idea, 1984, p. 187)

"We have so many gaps in the evolutionary history of life, gaps in such key areas as the origin of the multi-cellular organisms, the origin of the vertebrates, not to mention the origins of most invertebrate groups." (McGowan, C., In the Beginning... A Scientist Shows Why the Creationists are Wrong, Prometheus Books, 1984, p. 95)

"If living matter is not, then, caused by the interplay of atoms, natural forces and radiation, how has it come into being? I think, however, that we must go further than this and admit that the only acceptable explanation is creation. I know that this is anathema to physicists, as indeed it is to me, but we must not reject a theory that we do not like if the experimental evidence supports it." (H.J. Lipson, F.R.S. Professor of Physics, University of Manchester, UK, "A physicist looks at evolution" Physics Bulletin, 1980, vol 31, p. 138)

"Scientists have no proof that life was not the result of an act of creation, but they are driven by the nature of their profession to seek explanations for the origin of life that lie within the boundaries of natural law. They ask themselves, "How did life arise out of inanimate matter? And what is the probability of that happening?" And to their chagrin they have no clear-cut answer, because chemists have never succeeded in reproducing nature's experiments on the creation of life out of nonliving matter. Scientists do not know how that happened, and furthermore, they do not know the chance of its happening. Perhaps the chance is very small, and the appearance of life on a planet is an event of miraculously low probability. Perhaps life on the earth is unique in this Universe. No scientific evidence precludes that possibility." (Jastrow, Robert, The Enchanted Loom: Mind In the Universe, 1981, p. 19)

"Often a cold shudder has run through me, and I have asked myself whether I may have not devoted myself to a fantasy." (Charles Darwin, The Life and Letters of Charles Darwin, 1887, Vol. 2, p. 229)   END of Quotes

Also, this quoted from anointed-one site on Evolution: Quote – " For the sake of argument, let's assume something that was dead sprang to life in self-replicating form and spawned all life as we know it. And let's also imagine this was at least somehow represented in the fossil record. How would creatures be able to evolve and change into different ones? There are two ways: **Cross Breeding and Mutations**
First let's look at Cross Breeding. No one disputes the fact that horizontal variation (microevolution) exists. All species have a certain range of differences. But vertical transformation (macroevolution), where one kind of species transforms into another, is not allowed and does not occur. Dawkins tries to use the fact that snakes have a different number of vertebrae, microevolution, as proof that they evolved from another creature, macroevolution. This is a common, but false, argument.
Here is another amusing example from Dawkins, "You may find it hard to imagine an Amoeba turning into a man...but you do not find it hard to imagine an Amoeba turning into a slightly different kind of Amoeba. From this it is not hard to imagine it turning into a slightly different kind of slightly different kind of...and so on." While trying to postulate how macroevolution works, Dawkins uses dog breeding as an example, "Ah, but they are still dogs aren't they? They haven't turned into a different 'kind' of animal. Yes, if it comforts you to use words like that, you can call them all dogs." Boundaries between kinds are proven biological facts.
Centuries of experimental breeding provide convincing evidence against evolution. When abnormal lines are crossed, sterility is the result. For example, a horse and donkey can mate and produce a new animal, the mule. However, the mule is sterile and unable to procreate. The fact that hybrid offspring do not have the ability to reproduce is strong evidence against evolution. This is another example of how proven scientific facts confirm what the Bible says, each reproduces after its own kind.
Now on to Mutations. Many people believe mutations offer the best explanation for evolution. The following can be found on page 233 of *The Blind Watchmaker:* "The more 'macro' (a mutation) it is, the more likely it is to be deleterious, and the less likely it is to be incorporated in the evolution of the species. As a matter of fact, virtually all the mutations studied in genetic laboratories, are deleterious to the animals possessing them." For reasons I am not entirely sure of, Dawkins goes on to add, "Ironically, I've met people who think that this is an argument against Darwinism!"
Evolution is falling apart today. The theory of evolution is crumbling in the face of total lack of evidence. However, you might not hear about it because many people in this world simply find the only alternative to evolution, a divine creator, completely and totally unacceptable. If we were talking about any other subject, evolution would have been abandoned long ago, but atheists do not want to believe in God so they must believe in evolution by default, regardless of the fact that it does not have one leg to stand on. If evolution is a lie, then there is God. There is a moral standard. There is accountability for your actions." Unquote

Microevolution is a fact. All species have a range of variation and can have small changes in successive generations. Nobody should disagree with that. New species have been observed to form. In fact, *rapid* speciation is an important part of the creation model. But this speciation is within the "kind," and involves no new genetic information. However, macro-evolution (phyletic evolution) is a fraud with no basis in science. The main issue is not the size of the change but the direction. All observed change involves sorting and loss of genetic information.

Micro-macroevolution. These terms, which focus on "small" vs. "large" changes, distract from the key issue of information. That is, particles-to-people evolution requires changes that increase genetic information, but all we observe is sorting and loss of information. We have yet to see even a "micro" increase in information, although such changes should be frequent if evolution were true. Conversely, we do observe quite "macro" changes that involve no new information, e.g., when a control gene is switched on or off.

Creation – Intelligent Design

Even though the Bible isn't a science book, it is scientifically accurate. ClarifyingChristianity.com tells us how the Bible is like a science book. The verses used below are derived from the Holy Bible King James Version. http://www.clarifyingchristianity.com/science.shtml Used with Permission

Did you know the Bible frequently refers to the great number of stars in the heaven? (Astronomy)

Genesis 22:17 That in blessing I will bless thee, and in multiplying I will multiply thy seed as the stars of the heaven, and as the sand which is upon the sea shore; and thy seed shall possess the gate of his enemies; Jeremiah 33:22 As the host of heaven cannot be numbered, neither the sand of the sea measured: so will I multiply the seed of David my servant, and the Levites that minister unto me.

Even the scientists of today will admit they don't know how many stars there are.

The Bible also tells us that each star is unique as in 1 Corinthians 15:41 There is one glory of the sun, and another glory of the moon, and another glory of the stars: for one star differeth from another star in glory.

Even though they seem like points of light through a telescope and look alike to the naked eye, analysis of their light spectra reveals that each is unique and different.

The Bible says the stars cannot be counted: Used w/ permission from anointed-one.net

God brought Abraham outside and said, "Look now toward heaven and count the stars, if you are able to number them." Then God said to him, "So shall your descendants be." Genesis 15:5 written 1400 b.c. I will make the descendants of David, my servant, and the Levites who minister before me as countless as the stars of the sky and as measureless as the sand on the seashore. Jeremiah 33:22 written 600 b.c.

For centuries, astronomers thought they could count the precise number of stars in the sky. Like the number of grains of sand on the beach, the number of stars in the sky remains a great unknown. Eventually, scientists concluded the stars could not be numbered, just as the Bible always claimed. Discussing the number of stars in the Milky Way alone, astronomer William Keel claims there are "about as many as the number of hamburgers sold by McDonald's." Then he elaborates, "The usual way to determine the number of stars in the universe is to consider how many stars there are in the Milky Way, and then to multiply that number by our best estimate of the number of galaxies in the universe. This suggests there are probably about 400 billion stars in the Milky Way. The number of galaxies in the universe is an entirely different mathematical puzzle. Other estimates range from more than 200 billion stars in our galaxy to 3 thousand million billion stars (3 followed by 16 zeroes), in the universe. On a NASA web page, they state there are zillions of uncountable stars. End of info from anointed-one.net

The precision of movement in the universe is also describe in the Bible:
Jeremiah 31: 35, 36 35: Thus saith the LORD, which giveth the sun for a light by day, and the ordinances of the moon and of the stars for a light by night, which divideth the sea when the waves thereof roar; The LORD of hosts is his name: 36: If those ordinances depart from before me, saith the LORD, then the seed of Israel also shall cease from being a nation before me forever.

Here is how the Bible describes the suspension of the Earth in space.
Job 26:7 He stretcheth out the north over the empty place, and hangeth the earth upon nothing.

How does the Bible fit in with the subject of Physics?

Here is something that could not have been explained in 67 AD using our known scientific principles. The presence of nuclear processes is mentioned in the Bible, like those we associate with nuclear weaponry today. This is what Peter wrote in the following verse in 67 AD:

2 Peter 3:10 But the day of the Lord will come as a thief in the night; in the which the heavens shall pass away with a great noise, and the elements shall melt with fervent heat, the earth also and the works that are therein shall be burned up.

A device that uses electromagnetic waves is the television. The Bible does contain passages that describe something like television – something that allows all on earth to see a single event. (note: such passages typically refer to the end of time)

Matthew 24:30 And then shall appear the sign of the Son of man in heaven: and then shall all the tribes of the earth mourn, and they shall see the Son of man coming in the clouds of heaven with power and great glory. Revelation 11:9-11 - Re: 11:9 And they of the people and kindreds and tongues and nations shall see their dead bodies three days and an half, and shall not suffer their dead bodies to be put in graves. Re:11:10: And they that dwell upon the earth shall rejoice over them, and make merry, and shall send gifts one to another; because these two prophets tormented them that dwelt on the earth. Re:11:11: And after three days and an half the Spirit of life from God entered into them, and they stood upon their feet; and great fear fell upon them which saw them.

Biology in the Bible

Written prior to 1400 BC, the book of Leviticus describes the value of blood. The blood carries water and nourishment to every cell, maintains the body's temperature and removes the waste material of the body's cells. The blood also carries oxygen from the lungs throughout the body. William Harvey in 1616 discovered that blood circulation is the key factor in physical life. This confirms what the Bible revealed 3000 years earlier.

Leviticus 17:11 For the life of the flesh is in the blood: and I have given it to you upon the altar to make an atonement for your souls: for it is the blood that maketh an atonement for the soul.

Biogenesis in the Bible. The development of living organisms from other living organisms, along with the stability of each kind of living organism. Keep in mind, the phrase "after its kind" (according to its kind) occurs repeatedly. Which stresses the reproductive integrity of each kind of animal and plant.

Genesis 1:11, 12 And God said, Let the earth bring forth grass, the herb yielding seed, and the fruit tree yielding fruit after his kind, whose seed is in itself, upon the earth: and it was so. And the earth brought forth grass, and herb yielding seed after his kind, and the tree yielding fruit, whose seed was in itself, after his kind: and God saw that it was good. Genesis 1:21 And God created great whales, and every living creature that moveth, which the waters brought forth abundantly, after their kind, and every winged fowl after his kind: and God saw that it was good. Genesis 1:25 And God made the beast of the earth after his kind, and cattle after their kind, and every thing that creepeth upon the earth after his kind: and God saw that it was good.

We know this occurs today because all of these reproductive systems are programmed by their genetic codes.

What about the chemical nature of flesh? Let's see what the Bible says.

Genesis 2:7 And the LORD God formed man of the dust of the ground, and breathed into his nostrils the breath of life; and man became a living soul. Genesis 3:19 In the sweat of thy face shalt thou eat bread, till thou return unto the ground; for out of it wast thou taken: for dust thou art, and unto dust shalt thou return.

A person's **mental and spiritual health** is strongly associated with physical health, which is a proven fact. King Soloman, about 950 BC, wrote these statements and others, which are revealed in the Bible.

Proverbs 12:4 A virtuous woman is a crown to her husband: but she that maketh ashamed is as rottenness in his bones. Proverbs 14:30 A sound heart is the life of the flesh: but envy the rottenness of the bones. Proverbs 15:30 The light of the eyes rejoiceth the heart: and a good report maketh the bones fat. (make the bones healthy) Proverbs 16:24 Pleasant words are as an honeycomb, sweet to the soul, and health to the bones. Proverbs 17:22 A merry heart doeth good like a medicine: but a broken spirit drieth the bones.

Hydrology in the Bible.

Reasonably complete descriptions of the hydrologic cycle in the Bible.

Psalm 135:7 He causeth the vapours to ascend from the ends of the earth; he maketh lightnings for the rain; he bringeth the wind out of his treasuries.

The Bible also shows verses that contain the several phases of the hydrologic cycle- the worldwide processes of evaporation, translation aloft by atmospheric circulation, condensation with electrical discharges and precipitation.

Job:36:27: For he maketh small the drops of water: they pour down rain according to the vapour thereof: Job:36:28: Which the clouds do drop and distil upon man abundantly. Job:36:29: Also can any understand the spreadings of the clouds, or the noise of his tabernacle?

Here is a simple verse on how the Bible describes the recirculation of water. This verse has remarkable scientific insight on how drops of water, which eventually pour down as rain, first become vapor and then condense to tiny liquid water droplets in the clouds.

Ecclesiastes 1:7 All the rivers run into the sea; yet the sea is not full; unto the place from whence the rivers come, thither they return again. Isaiah 55:10 For as the rain cometh down, and the snow from heaven, and returneth not thither, but watereth the earth, and maketh it bring forth and bud, that it may give seed to the sower, and bread to the eater:

The surprising amount of water that can be held as condensation in the clouds are in Bible verses:

Job 26:8 He bindeth up the waters in his thick clouds; and the cloud is not rent under them. (the cloud is not rent under them means – they do not burst under their weight) Job 37:11 Also by watering he wearieth the thick cloud: he scattereth his bright cloud (means he loads the clouds with water and scatters his lightning through them) More than 3000 years ago, before their discovery by science, hydrothermal vents are described in two books of the Bible written before 1400 BC. Genesis 7:11 In the six hundredth year of Noah's life, in the second month, the seventeenth day of the month, the same day were all the fountains of the great deep broken up, and the windows of heaven were opened. Job 38:16 Hast thou entered into the springs of the sea? or hast thou walked in the search of the depth?

Meteorology in the Bible.

In this verse, the Bible describes the circulation of the atmosphere.

Ecclesiastes 1:6 The wind goeth toward the south, and turneth about unto the north; it whirleth about continually, and the wind returneth again according to his circuits.

The Bible also includes some principles of fluid dynamics.

Job 28:25 To make the weight for the winds; and he weigheth the waters by measure. (when he established the force of the wind and measured out the waters)

Only about 300 years ago, it was proven scientifically that air has weight. The relative weights of air and water are needed for the efficient functioning of the world's hydrologic cycle, which in turn sustains life on earth.

How does Anthropology fit in the Bible?

The Bible describes cave men, and we have paintings and other evidence that people inhabited caves in Job 30:5, 6. Job:30:5: They were driven forth from among men, (they cried after them as after a thief;) Job:30:6: To dwell in the clifts of the valleys, in caves of the earth, and in the rocks.

Note that these were not ape-men, but descendents of those who scattered from Babel. They were driven from the community by those tribes who competed successfully for the more desirable regions of the earth. Then for some reason they deteriorated mentally, physically and spiritually.

Geology in the Bible.
A verse from the Bible on the Earth's crust along with a comment on astronomy.

Jeremiah 31:37 Thus saith the LORD; If heaven above can be measured, and the foundations of the earth searched out beneath, I will also cast off all the seed of Israel for all that they have done, saith the LORD.

Although it is claimed by some scientists that they have measured the size of the universe, it is also interesting to note that every human attempt to drill through the earth's crust to the plastic mantle beneath has, so far, ended in failure.

Before people thought the earth was spherical (round), the Bible described the shape of the earth centuries ago.

Isaiah 40:22 It is he that sitteth upon the circle of the earth, and the inhabitants thereof are as grasshoppers; that stretcheth out the heavens as a curtain, and spreadeth them out as a tent to dwell in:

The word translated "circle" here is the Hebrew word chuwg which is also translated "circuit" or "compass" (depending on the context). That is, it indicates something spherical, rounded, or arched – not something that is flat or square.

What about Paleontology in the Bible?
Dinosaurs are referred to in several Bible books. The book of Job describes two dinosaurs. One is described in chapter 40 starting at verse 15, and the other in chapter 41 starting at verse 1. I believe you will agree that 1½ chapters about dinosaurs is a lot—since most people do not even realize that they are mentioned in the Bible.

As you can see, the Bible is very consistent with scientific facts. Scientists have discovered many proofs that confirm the Bible's accuracy in the last 100 years, and especially in the last 10 years.

Below Used with Permission from anointed-one.net

The Bible says springs and valleys exist on the ocean floor:

On the seventeenth day of the second month, on that day, all the springs of the great deep burst forth and the floodgates of the heavens were opened. Genesis 7:11 written 1400 b.c.

The springs of the deep and the floodgates of the heavens had been closed and the rain had stopped falling from the sky. Genesis 8:2 written 1400 b.c. Have you journeyed to the springs of the sea or walked in the recesses of the deep? Job 38:16 written 1900 b.c. The valleys of the sea were exposed and the foundations of the earth laid bare at the rebuke of the Lord. II Samuel 22:16 written 900 b.c.

Only in the last few decades did humanity have the technology to discover there are fresh water springs and deep valleys on the ocean floor. The Bible always knew.

Dinosaurs And The Bible http://www.clarifyingchristianity.com/dinos.shtml

Many of the common animals we know today are referred to in the Bible, such as dogs, sheep, insects, bears, wolves, and various kinds of birds, reptiles and rodents. It is interesting that this list doesn't contain three that we do not recognize. These three are leviathan, behemoth and tanniyn. We know dinosaurs existed, but many people may wonder why they aren't mentioned in the Bible? Well, actually they are.

Tanniyn does occur 28 times in the Bible and is normally translated "dragon" and it is also translated "serpent," "sea monster," "dinosaur," "great creature," and "reptile." Behemoth and Leviathan are relatively specific creatures, and so perhaps each was a single kind of animal. Tanniyn is a more general term, and it can be thought of as the original version of the word "dinosaur." The word "dinosaur" was originally coined in 1841, more than three thousand years *after* the Bible first referred to "Tanniyn."

Marilyn Oakley

It is interesting how we got these new names. In 1822, Mary Ann Mantell became the first person to discover and correctly identify a strange bone as part of a large, unknown reptile. Dr. Gideon Mantell, her husband, later named this creature an "Iguanodon." From that time on, these forgotten animals were given names chosen by the people who rediscovered them. The Bible was written between approximately 1450 BC and 95 AD, and of course does not include any of these names.

Yes, we realize that no animal living today matches the description of the behemoth and leviathan, but if you pick up a child's dinosaur book, you will note the several possible matches for each one.

Behemoth has the following attributes according to Job 40:15-24 –

Job:40:15: Behold now behemoth, which I made with thee; he eateth grass as an ox. Job:40:16: Lo now, his strength is in his loins, and [his force is in the navel of his belly. (what power in the muscles of his belly)] Job:40:17: He moveth his tail like a cedar: the sinews of his stones are wrapped together. (In Hebrew, this literally reads, "he lets hang his tail like a cedar.") Note - His tail was the size of a cedar tree, but some would like to say, perhaps the Bible meant elephant or hippo, but an elephant or hippos tail as we know, are much smaller and skinny, not huge like a cedar tree. The sinews of his stones are wrapped together mean his thighs are close-knit. Job:40:18: His bones are as strong pieces of brass; his bones are like bars of iron. Job:40:19: He is the chief of the ways of God: he that made him can make his sword to approach unto him. (He is the first of the ways of God, yet his maker can approach him with his sword.) Job:40:20: Surely the mountains bring him forth food, where all the beasts of the field play. Job:40:21: He lieth under the shady trees, in the covert of the reed, and fens. Job:40:22: The shady trees cover him with their shadow; the willows of the brook compass him about. Job:40:23: Behold, he drinketh up a river, and hasteth not: he trusteth that he can draw up Jordan into his mouth. Job:40:24: He taketh it with his eyes: his nose pierceth through snares. (Can anyone capture him by the eyes or trap him and pierce his nose?)

A key phrase is "He is the first of the ways of God." This phrase in the original Hebrew implied that behemoth was the largest animal created. Even though an elephant and hippo are big, they are less than 1/10 the size of a Brachiosaurus. Which is the largest complete dinosaur ever discovered. Thus making the Brachiosaurus therefore easily to be described as "the first of the ways of God."

So whatever a behemoth is, it is large, so it can be expected to be a large land animal whose bones are like beams of bronze, and so forth.

Some paleontologists have found fragmentary leg bones, ribs, or vertebrae that they suggest belong to "new" sauropods larger than Brachiosaurus. Examples of these include Amphicoelias, Argentinasaurus, Sauroposeidon, Seismosaurus, Supersaurus and Ultrasaurus. Currently there isn't enough evidence to really determine the size of any of these, and some paleontologists believe that they are merely large examples of known dinosaurs like Brachiosaurus or Diplodocus. In any case, only the "modern scientific name" of behemoth would change. The point would still remain that behemoth refers to a dinosaur, not a "modern animal" like an elephant or hippo.

Leviathan has the following attributes according to Job chapter 41, Psalm 104:25,26 and Isaiah 27:1. This is only a partial listing.

"No one is so fierce that he would dare stir him up." "Who can open the doors of his face, with his terrible teeth all around?" "His rows of scales are his pride, shut up tightly as with a seal; one is so near another that no air can come between them; they are joined one to another, they stick together and cannot be parted." "His sneezings flash forth light, and his eyes are like the eyelids of the morning. Out of his mouth go burning lights; sparks of fire shoot out. Smoke goes out of his nostrils, as from a boiling pot and burning rushes. His breath kindles coals, and a flame goes out of his mouth." "Though the sword reaches him, it cannot avail; nor does spear, dart, or javelin. He regards iron as straw, and bronze as rotten wood. The arrow cannot make him flee; slingstones become like stubble to him. Darts are regarded as straw; he laughs at the threat of javelins." "On earth there is nothing like him, which is made without fear." Leviathan "played" in the "great and wide sea" (a paraphrase of Psalm 104 verses 25 and 26). Leviathan is a "reptile that is in the sea." (Isaiah 27:1)

The word translated "reptile" here is the Hebrew word tanniyn. This shows that "Leviathan" was also a "tanniyn" (dragon). The Leviathan is ferocious and terrifying, unlike the behemoth who is huge. Many of the references refer to the sea, so the Leviathan is more than likely a sea creature. Some Bibles may refer to the Leviathan, as a crocodile or alligator, but neither of these is a sea creature. They do like the water but spend most of their time on land. The question "who can open the doors of his face ... " infers that nobody can open a Leviathan's jaws. While an alligator's jaw can't normally be forced open, a poke in the eye or a punch to their sensitive snout may startle them enough to release their grip.

Several verses describe these great scales. Iron is like straw and arrows cannot make it flee, is how the scales of the Leviathan is compared. Arrows can do a lot of damage to an alligator or crocodile, so we can eliminate that the Bible, when speaking of these large creatures, is not referring to gators or crocs. In reading Job 41:18-21, it tells us the Leviathan breathes fire, which totally eliminates the alligator and crocodile. It almost eliminates every living animal today, except the Bombardier Beetle. This beetle is a native of Central America. So if a Central America beetle can do this, then so could a Leviathan.

You may be wondering how the beetle can do this? Well, he has a nozzle in its hind end that acts like a little flamethrower. It can spray a high-temperature jet of gas, which is fueled by hydroquinones and hydrogen peroxide with oxidative enzymes.

Stories of fire-breathing dragons are in the history of every culture. Down the line, did someone ever make up a story about a fire-breathing bear, tiger or anything? No, because these dragon stories are based on truth. And only the dragons breathed fire. This makes it easy to imagine the Leviathan as a member of the dragon (tanniyn) family. Also, Isaiah 27:1 insinuates this connection.

Isaiah 27:1: In that day the LORD with his sore and great and strong sword shall punish leviathan the piercing serpent, even leviathan that crooked serpent; and he shall slay the dragon that is in the sea

Scientists haven't been able to guess the reason for empty passages in many fossil dinosaur skulls. They remain unexplained. So, wouldn't it make sense that some dinosaurs used these passages as "gas tanks" for the combustion mixture used to breathe fire. More on dinos and man http://www.creationists.org/wwd.html

There is a lot of recorded history that tells us about dinosaur and man. These items are art, literature and artifacts. How is it explained that the representation of recognizable species of dinosaurs--- by ancient men who supposedly had never seen one alive could have drawn or spoke of dinosaurs? Johann Johnston (1693-75), a doctor of medicine, published the celebrated: *De Serpentibus et Draconibus* in 1653. It featured many animals which are now extinct, all - according to the author, carefully drawn from living models. This in and of itself does not prove anything, however, their absence would be difficult to explain.

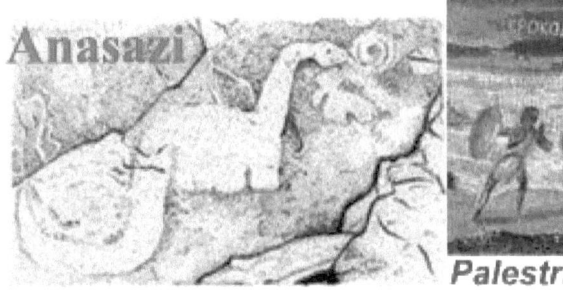

Palestrina Mosaic AD 100

Marilyn Oakley

What conditions have **the Flood** caused? Fossil remains of clams that were found in the closed position indicated they were buried alive, and have been found atop Mt. Everest. How about petrified trees that stand upright through multiple sedimentary layers, supposedly separated by millions of years? Think for a moment, these trees have been found upright. Now if these things had to wait for millions of years to fossilize, meaning waiting and waiting for a single layer to come along every so many thousands of years, how could they have preserved so well all the way to the top? By the time the top layer did cover the top of the trees completely, they would have decayed long before then. And also, how did they get in that upright position to begin with, so they could be covered over millions of years, layer by layer?

The Biblical story of the flood, also known, as Noah's Flood is the worldwide event that totally devastated the earth sometime between 2,000 and 3,000 BC. Earth's humanity had degenerated into such a cesspool, that God, in His sovereignty, decided to purge the earth. You can read about this in Genesis, chapters 6 through 9.

Noah was chosen because he was the man who "found grace in the eyes of the LORD" Genesis 6:8. God commanded him to build a massive ark, so that he and his family would survive the great flood. Along with every kind of animal, one male and one female. And two of every sort shall come onto Noah, meaning they would come to Noah.

Most of the fossil record doesn't represent a history of the Earth, but the order of burial during the Flood. This is what is found in the geological strata.

~~~

Many may ask, "Well, doesn't it take millions of years to fossilize?" The answer is no. It does not mean it took millions or even 1000's of years to turn into a fossil. A bone can become filled with minerals very quickly when the conditions and the materials are right. A quick burial, water in the right amounts and suitable minerals are the main ingredients.

Researchers have found that a big dinosaur bone may take hundreds of years to completely mineralize. Of course, it all depends on the burial conditions over the years. They have found that chicken bones can be replaced with minerals in as little as 5 years and up to 10 years.

Some of the plants from the Flood are not fossilized either. Large amounts of wood from trees that were growing at the same time as dinosaurs can be found in the Cretaceous clay in New Jersey. They are reserved and not turned to stone. The plants are not turned into rock and are just flattened and blackened. Since the Flood happened only a few thousand years ago, these types of discoveries are not surprising. Many dinosaurs' bones are still not turned completely into rock. More than half the fossil is still the original bone and not stone or rock. Red blood cells have been found in dinosaur's bones, how did these survive millions of years? Not possible.

Also reported in 2004, demonstrations showed that wood can petrify quickly. We are constantly fed the idea that fossils and rocks petrify over millions of years. You hear this in the media, newspapers, magazines and textbooks. But this idea is false. Recently, five Japanese scientists published more real examples of rapid petrification, confirming creationist claims. Scientists found that the naturally fallen wood in the overflow was hard and heavy because it was petrified with a mineral called silica. Yet the wood was less than 36 years old. So, millions of years are not needed to petrify wood. After seven years the wood had turned into stone. The evidence, both from scientists' laboratories and God's natural laboratory, shows that under the right chemical conditions wood can be rapidly petrified. Unfortunately, most people still think, and are led to believe, that fossilized wood buried in rock strata must have taken thousands, if not millions, of years to petrify. So these demonstrations of wood petrification can and does occur rapidly.

Well, one may ask also, where are all the people from before Noah's Ark? Why aren't there any human fossils from the Flood? The turbulence of water, even in a local flood, can be horrific. Especially when the fast moving current not only picks up mud and sand, but also large boulders. Under these conditions, human bodies would more than likely be thrown about and tend to be destroyed by the agitation and abrasion.

Even if human bodies had been buried in the later Flood sediments, destruction could still occur subsequently. Let me give an example; if ground waters permeating through the sediments, like sandstone, contain sufficient oxygen, then the oxygen would probably oxidize the organic molecules in the buried bodies and so destroy them. This could be regarded as a type of weathering. Likewise, chemically active ground waters could also be capable of dissolving human bones, removing all traces of buried people.

Another process that could destroy human bodies would be the intrusion of molten rock into the Flood sediments. Then through them to the surface to form volcanoes and lava flows. The heat from this process can be intense enough to melt rocks and recrystalize them. The heat often bakes these sediments as the molten rock rises through the sediments. So, then, again – chemical and mineral changes occur that obliterate many contained fossils. Many of these bodies were probably destroyed by the turbulence of the Flood waters as they were swept away.

Differential suspension is when the human body is immersed in water, tends to bloat, which in turn makes it lighter and it floats to the surface. This suspension would have made it hard to bury those bodies that survived the turbulence. These bodies floating would have made great carrion for any birds still flying around as they searched for a place to land and to eat food. Any marine carnivores still alive would also devour corpses. If they were not eaten and floated long enough, then they would either have to decompose or be battered to destruction on or in the waters before any burial took place.

Let's compare this to the 2004 tsunami in Southwest Asia. This tsunami was a shocking reminder of the speed with which water and other forces can eliminate all traces of bodies, even when we know where to look. Nearly 43,000 tsunami victims were never found according to the United Nation's Office of the Special Envoy for Tsunami Recovery.

Or, we can look at it like this. In Genesis 6:7 And the LORD said, I will destroy man whom I have created from the face of the earth; both man, and beast, and the creeping thing, and the fowls of the air; for it repenteth me that I have made them.

Perhaps the lack of such fossils is part of the fulfillment of this judgement? In reading Genesis 6:5-22, you read how God despised sin, how it corrupted the Earth and all in it. Perhaps God made sure there weren't any traces left at all.

The Apostle Peter takes up this theme in 2 Peter 3. He says that just as God created the world and judged the world the first time by the Flood, then so too He is going to keep His word and judge the world the second time by fire. Man therefore should take heed and make peace with his Creator while there is still time, before God comes again as Judge with sudden and swift judgment.

Do you think it's weird that **Eve was created from Adam**? Let me give you some facts on this as few people realize this. Dr. Robert Koontz had pointed out that the Genesis account of the creation of woman from man agrees with the modern knowledge of genetics, which was unknown to Moses.

In humans sex is determined by the two sex chromosomes. The female has in each body cell two X chromosomes, whereas the male has an X and a Y. Thus, if the female had been created first, it would not have been possible to create the first man from genetic material entirely related to the woman. This is because God in making Adam would have had to create Y chromosomes, for Eve had no Y chromosomes in her cells. As a consequence the resulting race would have been a hybrid race. But because man was created first, woman and man could be completely related to each other. Eve was the first clone! This unity of the race in Adam is theologically very important, for we all sinned in Adam and fell with him in his first transgression. The Redeemer of the fallen race, Jesus

Christ, receiving human nature by a miraculous conception in the womb of the Virgin Mary, became in a sense -- and said reverently -- a hybrid being, the God-Man. And all those who believe in Him are united with and in Him and receive a new nature, becoming the children of God by a spiritual rebirth. (See Romans 5 & 6 and I Corinthians 15.)

Cain's wife

Some of you who may have read the Bible, may question where Cain's wife came from? First, please let me cover some background information concerning the meaning of the Gospel. God didn't start by making a race of men, Adam was the first man, 1 Corinthians 15:14. The Bible does make it clear that only descendents of Adam can be saved. We are all descendents of Adam, with the exception of Eve, the first woman.

Because Adam sinned, we all sin and Romans 5 teach us this. Here is one scripture from Romans, Roman 5:12 Wherefore, as by one man sin entered into the world, and death by sin; and so death passed upon all men, for that all have sinned. Because of this, we are all separated from God since Adam brought sin and death into the world. The human race are all descendents of Adam.

Paul explains in 1 Corinthians 15, that God provided another Adam. The Son of God became a man, a perfect man. He is called the last Adam because he took the place of the first Adam. Because he was sinless, he was able to pay the penalty for sin; 1 Corinthians 15:21-22 For since by man came death, by man came also the resurrection of the dead. For as in Adam all die, even so in Christ shall all be made alive.

Jesus Christ suffered the penalty for sin, by dying on the cross. Acts 17:26 And hath made of one blood all nations of men for to dwell on all the face of the earth, and hath determined the times before appointed, and the bounds of their habitation; describes ALL human beings are sinners and all are related.

Cain as well as his wife was descendents of Adam, and descendents of Eve, the first woman. Cain was the first child of Adam & Eve. Abel and Seth, his brothers were part of the First generation. Adam and Eve did have other children, 33 sons and 23 daughters. Genesis 1:28 And God blessed them, and God said unto them, Be fruitful, and multiply, and replenish the earth, and subdue it: and have dominion over the fish of the sea, and over the fowl of the air, and over every living thing that moveth upon the earth. As you can see, they were commanded to be fruitful and multiply.

In order for future generations, brothers would have to marry sisters. Yes, I can hear you now, "Ugh! No way! Get real!" And so forth (lol). And yes, because of the law against brother-sister intermarriage. The law forbidding close relatives marrying weren't given until the time of Moses, Leviticus 18-20. Provided marriage was one man for one woman for life. There wasn't disobedience to God's law originally. This being before the time of Moses.

Of course today, brothers and sisters are not permitted to marry by law. It is true that children from such a marriage today, have a greater chance to be deformed. Genes today contain many mistakes because of sin, and these mistakes show up in a variety of ways. However this fact of present-day life did not apply to Adam and Eve. When they were created, they were perfect. Everything God made was "very good" Genesis 1:31. That means their genes were perfect, no mistakes. But when sin entered the world, Genesis 3:6, God cursed the world so that the perfect creation then began to degenerate, that is, to suffer death and decay. (Romans 8:22)

Over a long period of time, this degeneration would result in all sorts of mistakes occurring in the genetic material of living things. Since Cain and his siblings were First generation, they would have received virtually no imperfect genes from Adam or Eve. Because the effects of sin would have been minimal to start with.

I Was Blind But Now I See   Evolution - Creation

But by the time of Moses about 2500 years later, degeneration mistakes would have accumulated, making it necessary for God to bring in laws forbidding such marriages of brothers and sisters or close relatives.

You may or you may not be thinking, what does Cain's wife have to do with evolution and creation? Everything. As many will try to disprove the Bible, by asking, who is Cain's wife? Now, you have your answer, and as well as why it was possible in the beginning. When God created Adam, and then Eve, they were perfect. Nothing indecent about anything, until sin entered their perfect world. As in when sin entered their world, it was not until then they were ashamed of their nakedness, before that, it was normal and nothing to be ashamed of.

Dead Sea Scrolls

As you may or may not know, the Dead Sea Scrolls have been called the greatest archaeological find of the 20th century. They were found on the shores of the Dead Sea in the Judean Desert. You can read how the scrolls tell us about the development of Christianity, about the Hebrew Bible and the history of Judaism. For your free e-book which is a pdf file, please visit http://deadseascrolls.bib-arch.org/ and download. You will need adobe reader for this file, which you can get a free copy at www.adobe.com
For a free e-book of The Burial Of Jesus, please visit: http://jesustomb.bib-arch.org/

The Bible is made of many books written by people who lived in those times. The scrolls found are these books written by these people. The Bible is an authentic book. The manuscripts called the Dead Sea Scrolls represent over 900 separate writings. Some of them were written by the Jewish sectarians who formed the Qumran community. A great many were part of the wealth of literature circulating widely in Judea of the Second Temple period, and were brought to the site by the sectarians. These documents give us insights, not only into the workings of the Dead Sea sect itself, but also into the wider context and thought-world of Second Temple Judaism. From the scrolls on the Book Genesis, the flood is mentioned. In one scroll (4QGenb) they have only the following words preserved for Genesis 1:1, "In the beginning Go[] made [ ]." Fortunately, another scroll contains this part of Genesis 1:1, "In the begin[ ] God [ ] the heavens and the earth." This fragment includes material missing from 4QGenb. So when the preserved letters from the two scrolls are combined, the translation is evident: "In the beginning God made the heavens and the earth." (Abegg, Flint, and Ulrich, The Dead Sea Scrolls Bible, 3-5).

The article below available at http://evolution-facts.org/ Used with Permission Quote
SUMMARY: The Scientific Controversy Over Whether **Microevolution** Can Account For **Macroevolution**
© Center for Science and Culture/Discovery Institute, 1511 Third Avenue, Suite 808, Seattle, WA 98101

When Charles Darwin published *The Origin of Species* in 1859, it was already known that existing species can change over time. This is the basis of artificial breeding, which had been practiced for thousands of years. Darwin and his contemporaries were also familiar enough with the fossil record to know that major changes in living things had occurred over geological time. Darwin's theory was that a process analogous to artificial breeding also occurs in nature; he called that process natural selection. Darwin's theory was also that changes in existing species due primarily to natural selection could, if given enough time, produce the major changes we see in the fossil record.

After Darwin, the first phenomenon (changes within an existing species or gene pool) was named "microevolution." There is abundant evidence that changes can occur within existing species, both domestic and wild, so microevolution is uncontroversial. The second phenomenon (large-scale changes over geological time) was named "macroevolution," and Darwin's theory that the processes of the former can account for the latter was controversial right from the start. Many biologists during and after Darwin's lifetime have questioned whether the natural counterpart of domestic breeding could do what domestic breeding has never done -- namely, produce new species, organs, and body plans. In the first few decades of the twentieth century, skepticism over this aspect of evolution was so strong that Darwin's theory went into eclipse. (See Chapter 9 of Peter Bowler's *Evolution: The History of an Idea*, University of California Press, revised edition, 1989).

In the 1930s, "neo-Darwinists" proposed that genetic mutations (of which Darwin was unaware) could solve the problem. Although the vast majority of mutations are harmful (and thus cannot be favored by natural selection), in rare instances one may benefit an organism. For example, genetic mutations account for some cases of antibiotic resistance in bacteria; if an organism is in the presence of the antibiotic, such a mutation is beneficial. All known beneficial mutations, however, affect only an organism's biochemistry; Darwinian evolution requires large-scale changes in morphology, or anatomy. Midway through the twentieth century, some Darwinian geneticists suggested that occasional "macromutations" might produce the large-scale morphological changes needed by Darwin's theory. Unfortunately, all known morphological mutations are harmful, and the larger their effects the more harmful they are. Scientific critics of macromutations took to calling this the "hopeful monster" hypothesis. (See Chapter 12 of Bowler's book.)

The scientific controversy over whether processes observable within existing species and gene pools (microevolution) can account for large-scale changes over geological time (macroevolution) continues to this day. Here are a few examples of peerreviewed scientific articles that have referred to it just in the last few years:

David L. Stern, "Perspective: Evolutionary Developmental Biology and the Problem of Variation," *Evolution* 54 (2000): 1079-1091.

"One of the oldest problems in evolutionary biology remains largely unsolved...
Historically, the neo-Darwinian synthesizers stressed the predominance of micromutations in evolution, whereas others noted the similarities between some dramatic mutations and evolutionary transitions to argue for macromutationism."

Robert L. Carroll, "Towards a new evolutionary synthesis," *Trends in Ecology and Evolution,* 15 (January, 2000): 27.

"Large-scale evolutionary phenomena cannot be understood solely on the basis of extrapolation from processes observed at the level of modern populations and species."

Andrew M. Simons, "The continuity of microevolution and macroevolution,"
Journal of Evolutionary Biology 15 (2002): 688-701.

"A persistent debate in evolutionary biology is one over the continuity of microevolution and macroevolution -- whether macroevolutionary trends are governed by the principles of microevolution."

It should be noted that all of the scientists quoted above are believers in Darwinian evolution, and that all of them think the controversy will eventually be resolved within the framework of that theory. Stern, for example, believes that new developmental studies of gene function will provide "the current missing link." (p. 1079) The important point here is that the controversy has not yet been resolved, precisely because the evidence needed to resolve it is still lacking. It is important for students to know what the evidence does or does not show -- not just what some scientists hope the evidence will eventually show.

Since the controversy over microevolution and macroevolution is at the heart of Darwin's theory, and since evolutionary theory is so influential in modern biology, it is a disservice to students for biology curricula to ignore the controversy entirely. Furthermore, since the scientific evidence needed to settle the controversy is still lacking, it is inaccurate to give students the impression that the controversy has been resolved and that all scientists have reached a consensus on the issue. Unquote

Survival Of The Fakest - SCIENCE NOW KNOWS THAT MANY OF THE PILLARS OF DARWINIAN THEORY ARE EITHER FALSE OR MISLEADING. YET BIOLOGY TEXTS CONTINUE TO PRESENT THEM AS FACTUAL EVIDENCE OF EVOLUTION. WHAT DOES THIS IMPLY ABOUT THEIR SCIENTIFIC STANDARDS? -- JONATHAN WELLS

From the www.evolution-facts.org Used with Permission
Article by Jonathan Gray
Quote Unnatural Graveyards
Well, let your mind go wild on this!

A DESPERATE DASH TO SAFETY
Can you just imagine it! The terror-stricken dash by man and animal alike to reach safety - only to be sucked under by the relentless force of advancing waters. The agonized cries of drowning men and beasts piercing the air.

UNNATURAL GRAVEYARDS
Gripped by the same terror, wild beasts and tame struggle together to higher grounds. Some of the people bind their children and themselves upon powerful animals, knowing that these will climb to the highest peaks to escape the rising waters. Some fasten themselves to tall trees on the hills or mountains, but the trees are uprooted and hurled into the billows.

As the waters rise higher, the people flee for refuge to the loftiest heights. In amphitheatres in the hills, they find themselves trapped. In great numbers they throng together, pushing into caves, swarming over the ground in front. Until the waters rise and cover them.

Strong animals, without a sign of degeneration, come to an end. Yes, worldwide, the traditions of the human race recount this terrifying disaster. They recall that it was accompanied by enormous universal rains, violent waterspouts, earthquakes and hurricane winds.

THE FIT DO NOT SURVIVE
This is not the survival of the fittest. Fit and unfit, and mostly fit, old and young with sharp teeth, with strong muscles, with fleet legs, with plenty of food around, all perish. The earth is even at this moment convulsing, opening up fissures to swallow many of them, as they collect on the tops of these hills. Then the huge waves smash over them large rocks and debris, until their bones are crushed and smashed. Here, often thousands of feet up, they are washed into crevices and held tight.

WHAT YOUR TEACHER DIDN'T TELL YOU
There is a most crucial circumstance concerning the earth's strata and the fossils that is not generally disclosed to the public, and which many geologists apparently do not recognise. On every continent, and in numerous places on each, are vast "fossil graveyards", where masses of flora and fauna have been swept to a sudden death in their millions. These areas are often packed with both land and sea creatures from different habitats and even different climatic zones - all mixed and buried together in a completely unnatural way. There is evidence that a great disaster took place, in which creatures of all types perished together - mostly fit, young and old, with fleet legs, strong muscles and sharp teeth.

And with plenty of food around. Artifacts of man are found among them. They all died together, suddenly and violently, high up on hills and mountains.

Marilyn Oakley

What did it? Come with me on a global tour. You'll see what I mean.

In France, numerous clefts crammed to overflowing with animal bones have been found. Along with them are human remains. One could mention Mount Genay, near Semur in Burgundy - 1,430 feet high. Here, capped by a breccia (a cemented mass of stone fragments), is a fissure filled with mixed bones of numerous animals.

Near Chalon-sur-Saone, between Dijon and Lyons, stands an isolated hill, flat-topped Mont de Sautenay. It rises 1,030 feet above the plain. Near the summit is a fissure crammed with bones. The bones are unweathered and un-gnawed.

I ask you, Why should so many wolves, bears, horses and oxen have ascended a hill isolated on all sides? These broken and splintered bones are evidently not those of animals devoured by beasts of prey; nor have they been broken by man. The state of preservation of the bones indicates that the animals perished in the same period of time. It appears that all these animals had fled there to escape rising waters.

Caves and fissures on the Cote d'Azur have yielded mixed land and sea remains - bones of lions, rhinoceros, hyenas, macao monkeys, elephants and whales. All together. In Britain (for example, Yorkshire, Plymouth, Devonshire and Wales) and Eire, are similar mixtures.

What could have brought, together in quick succession, all these animals and plants, from the tundra of the Arctic Circle and from the jungle of the tropics. from lands of many latitudes and altitudes, from freshwater lakes and rivers, and from the salt seas of the north and south? There is evidence of submersion to a depth of not less than about 1,000 feet.

In Germany, coal beds contain a complete mixture of plants, insects and animals from all climatic zones. Their muscles and skin are perfectly preserved.

Although leaves are preserved in fresh condition (retaining their fine fibre and green colour) and insects their membranes and colours perfectly preserved, ALL ARE VIOLENTLY TORN INTO PIECES.

Were they carried there by onrushing water - from all parts of the world???

Numerous crevices down to 290 feet deep on the Rock of Gibraltar are filled with bones of the wolf, bear, lynx, hare, ibex, rabbit, horse, panther, rhinoceros, ox, wild boar, deer and other animals.

The bones are neither worn, rolled, nor gnawed. They are splintered and broken. And found together with land and marine shells - as well as coral.

Tell me, what but a great and common danger, such as a colossal flood, could have driven together the animals of the plains and of the crags and caves?

A handful of bone samples still in the Natural History Museum in London confirm their "freshness", indicating that they had been entombed with the flesh still on them - and not eaten by predators.

These show a consistently unrolled and fresh appearance. Flood, not glacial action, is indicated. We could cite similar evidence from Sicily - where bones of hippopotamus, deer, ox and elephant were mixed together in wild confusion, again without sign of weathering or gnawing.

The fact that bones of animals of all ages were piled together, indicates that the catastrophe was SUDDEN.

On the island of Malta are caves crammed full of hares, lemurs, tigers, rhinoceros, elephants - and birds.

Giant swans (twice as large as any swans now on earth) and huge sea birds are crammed in with the animals.

Now Malta is an island 170 miles (270 km) from the nearest other land (Sicily).

And please note, these birds are sea birds.

I Was Blind But Now I See   Evolution - Creation

Had this been a mere local flood, these giant swans would have taken off for Sicily and been there in half an hour.

These sea birds would never have been drowned - they're sea birds! So what happened?

You've guessed it! These birds kept flying until they could fly no more!!!

The same mixture of LAND animals and SEA shells is found in caves and fissures from Western Europe to Russia - and in Cyprus, Greece, Lebanon and elsewhere. The bones are broken into innumerable fragments. They are still "fresh". Artifacts of man are found among them. Fissures are in rocks on top of isolated hills. Often, birds and animals are MIXED TOGETHER WITH trees and vegetation.

THE CAUSE?

QUESTION: "COULD THEY HAVE FALLEN IN ALIVE, OR BEEN BURIED THERE"

The answer is no, for there are no complete skeletons.

QUESTION: "COULD THEY HAVE BEEN BROUGHT THERE BY STREAMS?"

Again, no, for there are no signs of them having been rolled.

QUESTION: "COULD THEY HAVE BEEN SLAUGHTERED BY RUTHLESS MEN?"

No, there could have been no way to bury them in such numbers before other animals gnawed them or the weather affected them.

Considering the number of the remains found, thrown together in great heaps, mixed in great confusion, large and small animals, grass-eating and flesh-eating animals and birds, all in one pile, buried in alluvial deposits, intermingled with remains of plants and trees, seashells and fish, such an idea becomes absurd.

QUESTION: "DID SOME DISEASE STRIKE THEM DOWN?"

All animals at the same time and in every part of the world? Land animals and deep sea creatures, all mixed together? Of course not.

Certainly animals stricken by sudden disease would be too weak to ascend to the hilltops!

The bones could not have been exposed to weather for long, for none of them shows marks of weathering. That water deposited them is indicated by the very general cementing together of the bones by calcite.

Moreover, these bone-filled chasms are usually found on isolated hills of considerable height - places on which we might expect animals to gather in seeking safety from an approaching flood.

This scene is enacted on hundreds of thousands of hills in every part of the earth. While panicked men, women and animals were rushing to the highest points in Europe, the very same scene was being enacted ALL OVER THE WORLD. From North America (California, Maryland, Utah, Colorado, Wyoming, New Mexico and Canada, for example), the story is the same.

Human remains are mixed chaotically with those of land and sea animals - just as in South America, Australia, Africa, Siberia, India, China, Korea, Burma. In fact, all over Asia.

WHAT COULD DO THIS, BUT WATER?

Clearly, nothing but a flood would have driven such strange mixtures of animals - animals that don't normally live together - into caves and crevices in all parts of the world. And buried them together.

The evidence everywhere strongly suggests that they were drowned en masse by violent water action. This bone-cave phenomenon is of world-wide extent.

Was this a global Flood? Of course it was.

Here is a fact of utmost importance: Creatures of every kind died in great numbers and were buried almost instantly. Each of these animals, by the unbelievable uniformitarian explanation, fell into their graveyards by accident - one at a time!

However, the facts reveal not normal, slow processes, but unusual transportation and rapid burial mechanisms.

DINOSAURS LIKEWISE DROWNED

Yes, big as they were, dinosaurs suffered the same watery fate - followed by rapid burial. This is evidence that some scientists prefer to ignore. Great dinosaur beds have been excavated in Alberta, Belgium, New Mexico, Tanzania, Spitzbergen and many other places.

All over the world, paleontologists have found caches of fossilized dinosaurs that were buried instantly in a catastrophic movement of water.

Fascinating, don't you think?

And yet, we have hardly scratched the surface. If you would like to discover more, here's where you can go: http://www.beforeus.com/second.php

Warm regards, Jonathan Gray info@archaeologyanswers.com

International explorer, archaeologist and author Jonathan Gray has traveled the world to gather data on ancient mysteries. He has penetrated some largely unexplored areas, including parts of the Amazon headwaters. The author has also led expeditions to the bottom of the sea and to remote mountain and desert regions of the world. He lectures internationally. Unquote

Has **human culture** evolved through successive stages from cave dwellers to nomadic hunters to farm village dwellers and at last, to builders of great cities-states? The answer to this is no. The evidence from archaeology shows the sudden appearance of the advanced Sumerian civilization without any signs of slow evolution upward from cave men. These observed facts really fit what the Bible says. When the Sumerian people appeared, they brought with them their art, the potter's wheel, as well as writing, religion and government, which all were in a highly developed state. Archaeology supports the biblical model for the origins of ancient civilization.

As you know, the human fossils have been found in the wrong strata to even support the evolution theory. For the most part, fossils not fitting this theory are ignored or explained away. Fossil remains that are the same order to the same order as modern man have been ignored or are no longer reported. British Anthropologist Sir Arthur Keith describes fossil-man finds in great detail in his book, *The Antiquity Of Man*. He states in his book that scientists would have readily accepted these finds if it weren't for the fact that these fossils contradicted the accepted theory of human evolution. Because of these fossils locations in the strata, the theory goes down the drain.

God created the different languages along with the different nationalities; Genesis 11:9 Therefore is the name of it called Babel; because the LORD did there confound the language of all the earth: and from thence did the LORD scatter them abroad upon the face of all the earth.

Below notes taken from various pages of "How did all the different 'races' arise (from Noah's family)?"

The Bible teaches us that God has 'made of one blood all nations of men', Acts 17:26 And hath made of one blood all nations of men for to dwell on all the face of the earth, and hath determined the times before appointed, and the bounds of their habitation.

According to the Bible, all humans on Earth today descended from Noah and his wife, his three sons and their wives, and before that from Adam and Eve Genesis 1–11. There is really only one race, the human race. From Genesis 11, it is understood that up to this time there was only one language. God judged the people's disobedience by imposing different languages, so that they could not work together against God. The confusion forced the people to scatter over the Earth as God intended. So all the people groups—

black Africans, Indo-Europeans, Mongolians, and others have come into existence since Babel.

The first created man, Adam, from whom all other humans are descended, was created with the best possible combination of genes or skin color, for example. A long time after Creation, a worldwide Flood destroyed all humans except a man called Noah, his wife, his three sons, and their wives. This Flood greatly changed the environment. Afterwards, God commanded the survivors to multiply and fill the Earth, Genesis 9:1. About a hundred years later, people chose to disobey God and to remain united in building a city, with the Tower of Babel as the focal point of rebellious worship.

Scripture distinguishes people by tribal or national groupings, not by skin color or physical features. The Bible tells us how the population that descended from Noah's family had one language and by living in one place was disobeying God's command to 'fill the earth' (Gen. 9:1, 11:4). God confused their language, causing a break-up of the population into smaller groups, which scattered over the Earth (Gen. 11:8–9). Modern genetics shows how, following such a break-up of a population, variations in skin color, for example, can develop in only a few generations. There is good evidence that the various people groups we have today have not been separated for huge periods of time.

To download a copy to read further on this. 'How did all the different 'races' arise' http://www.marilynoakley.com/RacesFromBible.pdf

~~~

"Utterly Impossible" Used with Permission from evolution-facts.org

Chapter 24: **Utterly Impossible**

Things Evolution could Never Invent
1 - FACTS WHICH CANNOT BE DENIED

It is commonly said that evolution and Creation are both theories. A "theory" has no definite proof in its support, only some evidence favoring it. In this book, we have found that evolution has no evidence supporting it and a ton of facts which destroy it.

But Creation is different. It has a mammoth number of facts from the natural world supporting it. And those facts do not fit any other possible explanation of earth or galactic origin. Regardless of what the evolutionists may claim, Creation is not a theory; it is a proven scientific fact.

To fill space at the end of the chapters in this book, a sampling of facts from the natural world have been included; each of which could only be explained by Creation. (Most are listed at the beginning of the index on page 980.)

Here are three more. As you read them, be open-minded and think. Accept the reality of the situation. Our world was made by a super-powerful, massively intelligent Creator. The world did not make itself.

ANATOMY OF A WORKER BEE

(1) Compound eyes able to analyze polarized light for navigation and flower recognition. (2) Three additional eyes for navigation. (3) Two antennae for smell and touch. (4) Grooves on front legs to clean antennae. (5) Tube-like proboscis to suck in nectar and water. When not in use, it curls back under the head. (6) Two jars (mandibles) to hold, crush, and form wax. (7) Honey tank for temporary storage of nectar. (8) Enzymes in honey tank which will ultimately change that nectar into honey. (9) Glands in abdomen produce beeswax, which is secreted as scales on rear body. (10) Five segmented legs which can turn in any needed direction. (11) Pronged claws, on each foot, to cling to flowers. (12) Glands in head make royal jelly. (13) Glands in body make glue. (14) Hairs on head, thorax, and legs to collect pollen. (15) Pollen baskets on rear legs to collect

pollen. (16) Several different structures to collect pollen. (17) Spurs to pack it down. (18) Row of hooks on trailing edges of front wings, which, hooking to rear wings in flight, provide better flying power. (19) Barbed poison sting, to defend the bee and the hive. (20) An enormous library of inherited knowledge regarding: how to grow up; make hives and cells; nurse infants; aid queen bee; analyze, locate, and impart information on how to find the flowers; navigate by polarized and other light; collect materials in the field; guard the hive; detect and overcome enemies;—and lots more!

How can a honeycomb have walls which are only 1/350th an inch [.007 cm] thick, yet be able to support 30 times their own weight?

How can a strong, healthy colony have 50,000 to 60,000 bees—yet all are able to work together at a great variety of tasks without any instructors or supervisors?

How can a honeybee identify a flavor as sweet, sour, salty, or bitter? How can it correctly identify a flower species and only visit that species on each trip into the field—while passing up tasty opportunities of other species that it finds en route?

All these mysteries and more are found in the life of the bee. A honeybee averages 14 miles [22.5 km] per hour in flight, yet collects enough nectar in its lifetime to make about 1/10th of a pound [.045kg] of honey. In order to make a pound of honey, a bee living close to clover fields would have to travel 13,000 miles [20,920 km], or 4 times the distance from New York City to San Francisco!

With all this high-tech equipment on each bee, surely it must have taken countless ages for the little bee to evolve every part of it. Yet, not long ago, a very ancient bee was found encased in amber. Analyzing it, scientists decided that, although it dated back to the beginning of flowering plants, it was just like modern bees! So, as far back in the past as we can go, we find that bees are just like bees today!

PORTRAIT FROG

At random, we will select one of several hundred examples we could cite.

The South American false-eyed frog is an interesting creature. Generally about 3 inches [7.62 cm] long, it is brown, black, blue, gray, and white! Drops of each color are on its skin, and it can suddenly change from one of these colors to the others, simply by masking out certain color spots.

The change-color effect that this frog regularly produces is totally amazing, and completely unexplainable by any kind of evolutionary theory.

The frog will be sitting in the jungle minding its own business, when an enemy, such as a snake or rat, will come along.

Instantly, that frog will jump and turn around, so that its back is now facing the intruder. In that same instance, the frog changed its colors!

Now the enemy sees a big head, nose, mouth, and two black and blues eyes!

All this looks so real—with even a black pupil with a blue iris around it. Yet the frog cannot see any of this, for the very highly intelligently designed markings are on its back!

The normal sitting position of this frog is head high and back low. But when the predator comes, he quickly turns around, so his back faces the predator! In addition, the frog puts his head low to the ground and his hind parts high. In this position, to the enemy viewing him, he appears to be a large rat's head! In just the right location is that face and eyes staring at you!

The frog's hind legs are tucked away together underneath his eyes—and they look like a large mouth! As he moves his hind legs, the mouth appears to move! The part of the frog's body that once was a tadpole's tail—now looks like a perfectly formed nose; and it is just at the right location!

I Was Blind But Now I See   Evolution - Creation

To the side of the fake face, there appear long claws! These are the frog's toes! As the frog tucks his legs to the sides of his body, he purposely lifts up two toes from each hind foot—and curls them out, so they will look like a couple of weird hooks.

And the frog does all this in one second!

At this, the predator leaves, feeling quite defeated. But that which it left behind is a tasty, defenseless, weak frog which can turn around quickly, but cannot hop away very fast.

The frog will never see that face on itself, so it did not put the face there. Someone very intelligent put that face there! And the face was put there by being programmed into its genes. Well, there it is. And it is truly incredible.

How could that small, ignorant frog, with hardly enough brains to cover your little fingernail do that? Could that frog possibly be intelligent enough to draw a portrait on the ground beneath it? No, it could not. Could it do it in living color? No!

Then how could it do it on its own back? There is no human being in the world smart enough—unaided and without mirrors—to draw anything worthwhile on his own back. How then could a frog do it?

It cannot see its back, just as you cannot see yours. The task is an impossible one. And, to make matters more impossible, it does it without hands! Could you, unaided by devices or others, accurately draw a picture on your back? No. Could you do it simply by making colors to emerge on the skin? A thousand times, No.

"Portrait frog"! This is the motion-picture frog! And the entire process occurs on its back, where it will never see what is happening! And it would not have the brains to design or prepare this full-color, action pantomime even if it could see it.

Someone will comment that frogs learn this by watching the backs of other frogs. But the picture is only formed amid the desperate crisis of encountering an enemy about to leap upon it. Only the enemy sees the picture; at no other time is the picture formed.

All scientists will agree that this frog does not do these things because of intelligence, but as a result of coding within its DNA. How did that coding get there? It requires intelligence to produce a code. Random codes are meaningless and designs never arise though random activity. They require intelligent planning. Genetic codes within living creatures are the most complicated of humans to devise and fabricate.

The facts are clear. God made that frog, and He made all other living creatures also. Only His careful thought could produce and implant those codes and the physical systems they call for. There can be no other answer.

THE PALOLO WORM

As our third and last example, we will tell you about a lowly blind worm who lives all but a few days of his life in the black depths of the ocean. The palolo worm is as incredible as many other creatures. Randomness could never produce this. Neither natural selection (the proper name for it is "random accidents") nor mutations could invent the palolo worm.

Palolo worms live in coral reefs off the Samoan and Fijian Islands in the south Pacific. Twice a year, with astounding regularity, half of this worm develops into another animal with its own set of eyes, floats to the surface on an exact two days in one or the other of two months in the year, and then spawns!

Yet these worms live in total darkness and isolation in coral holes deep within the ocean, have no means of communicating with one another, nor of knowing time—not even whether it is night or day! How can they know when it is time to break apart for the spawning season? Here is the story of the Palolo worm:

The palolo worm (Eunice virdis) measures about 16 inches [41cm] long. It lives in billions in the coral reefs of Fiji and Samoa in the Southwestern Pacific. The head of an individual worm has several sensory tentacles and teeth in its pharynx. Males are reddish-brown and females are bluish-green. These worms go down into the deep coral atolls and riddle it with their tiny, isolated tubes. They also burrow under rocks and into crevices. Once settled into their homes, these creatures catch passing food—small polyps—with their "tails" while their heads are buried inside the coral or between rock.

The body of one of these worms is divided into segments, like an earthworm's body; and each contains a set of the organs necessary for life. But reproductive glands only develop in rear segments.

As the breeding season nears, the "brain" of the little worm, inside the coral, decides that the time has come for action. The back half of the palolo worm alters drastically. Muscles and other internal organs in each segment grow rapidly. Then the pololo worm partially backs out of its tunnel and the outer half breaks off. By that time, the other half has grown its own set of eyes! Once separated from the rest of the worm, the broken-off half swims to the surface. (Down below in the coral, the "other half" grows a new back half and continues on with life.)

On reaching the surface, the free-swimming halves break open; their eggs and sperm float in the water; and fertilization occurs. The empty skins sink to the bottom, devoured by fish as they go. Soon, free-swimming larvae develop and, becoming full grown palolo worms, they sink deep into the ocean and burrow into the reefs.

We have here a creature which stays at home while sending off part of itself to a distant location to produce offspring. That is astounding enough. But the most amazing part is the clockwork involved in all this! The success of this technique depends upon timing. If the worms are to achieve cross-fertilization, they all must detach their hind parts simultaneously. So all those worm segments are released at exactly the same time each year!

Swarming occurs at exactly the neap tides which occur in October and November. (Some of the spawning occurs in October, but mostly in November.) It occurs at dawn on the day before and the day on which the moon is in its last quarter.

Suddenly, all the half-worms are released into the ocean. Swimming to the surface and bursting open, the sea briefly becomes a writhing mass of billions of worms and is milky with eggs and sperm. The timing is exquisite.

People living in Samoa and Fiji watch closely as these dates approach. When the worms come to the surface, boats are sent out to catch vast numbers of them. They are shared around; festivals are held, and the worms are eaten raw or cooked. In Fiji, the Scarlet aloals and the seasea flowers both bloom. This is the signal that the worms are about to rise to the surface! Then, each morning, the nationals watch for the sun to be on the horizon just as day breaks. Ten days after this—exactly ten days—the palolo worms will spawn. The first swarm is called Mbalolo lailai (little palolo), and the second is Mbalolo levu (large palolo). On the island of Savaii, the swarming is predicted by the land crabs. Exactly three days before the palolo worms come to the surface, all the land crabs on the island mass migrate down to the sea to spawn.

Throughout those islands, the nationals know to arise early on the right day. An hour or so before dawn, some will begin wading in darkness, searching the water with torches for evidence of what will begin within an hour. Even before the night pales into dawn, green wriggling strings will begin to appear in the black water. Flashlights reveal them, vertically wriggling upward toward the surface. Shouts are raised; the palolo worms have been seen! People who have been sleeping on the beaches awake. Gathering up their nets, scoops, and pails, they wade out into the water. Dawn quickly follows, and now the number of worms increases astronomically! Billions of worms have risen and are floating on large expanses of the ocean's surface. The sea actually becomes curdled several inches deep with these tiny creatures;—yet only a half hour before there were hardly any, and absolutely none before that for nearly a year. The people ladle them into buckets, as large fish swim in and excitedly take their share.

People and fish must work fast; an hour before there were none,—and already the worms are breaking to pieces! As their thin body walls rupture, the eggs and sperms come out and give a milky hue to the blue-green ocean. Quickly, the empty worm bodies fall downward into the ocean and disappear.

Within half-an-hour after the worms first appear, they are gone, —and only eggs and sperm remain.

Scientists have tried to figure out how the palolo worm calculates the time of spawning so accurately. But there is just no answer. The worms cannot watch the phases of the moon from their burrows. They are too far down in the ocean to see light or darkness or note the flow of the tides. The only solution appears to be some kind of internal "clock"!

But wait, how can that be? An internal clock would require that the action be triggered every 365 days, but this cannot be; since the moon's movements are not synchronized with our day-night cycle, the movements of the sun, nor with our calendar.

As a result, the moon's third quarter in October arrives ten or eleven days earlier each year until it slips back a month. Nor can it be that the worms in their holes are somehow able to judge the phase of the moon by the light; for they spawn whether the sky is clear or completely overcast.

Well then, it must be that the worms send signals to each other through the water! But that cannot be; for the palolo worms on the reefs of Samoa split apart at exactly the same time as the worms at Fiji—which are 600 miles away! If some kind of signal could indeed be sent over such a vast stretch of ocean, it would take weeks to arrive.

Indeed, the timing appears to have been pre-decided for the worm. There is no celestial or oceanic logic to it. The Pacific palolo spawns at the beginning of the third quarter in October or November; whereas the Atlantic palolo—near Bermuda and the West Indies—also spawns at the third quarter, but always in June or July instead of October! (Far away from both, a third pololo worm also spawns yearly at the beginning of the third quarter in October or November.)

At any rate, the advantages are obvious. All the eggs and sperm are together for a few hours, and a new generation is produced. Some other sedentary creatures also reproduce within narrowed time limits. This includes oysters, sea urchins, and a variety of other marine animals. But, with the exception of the California coast grunion, none do it within such narrowed, exacting time limits as the palolo worm.

Our Creator made the honeybee, the portrait frog, the palolo worm—and everything else in our world. May we acknowledge Him, honor Him, and serve Him all the days of our life. He deserves our truest, our deepest worship and service; for He is our Creator and our God.

2 - CONCLUSION

Few men in Europe have tried to eradicate the Bible and the knowledge of God from the minds of the people as did the French infidel, Voltaire. The Christian physician who attended Voltaire, during his last illness, later wrote about the experience:

"When I compare the death of a righteous man, which is like the close of a beautiful day, with that of Voltaire, I see the difference between bright, serene weather and a black thunderstorm. It was my lot that this man should die under my hands. Often did I tell him the truth. 'Yes, my friend,' he would often say to me, 'you are the only one who has given me good advice. Had I but followed it, I should not be in the horrible condition in which I now am. I have swallowed nothing but smoke. I have intoxicated myself with the incense that turned my head. You can do nothing for me. Send me an insane doctor! Have compassion on me—I am mad!'

Marilyn Oakley

"I cannot think of it without shuddering. As soon as he saw that all the means he had employed to increase his strength had just the opposite effect, death was constantly before his eyes. From this moment, madness took possession of his soul. He expired under the torments of the furies

---

An American tourist, in France, went to the hotel keeper to pay his bill. The French hotel keeper said, "Don't you want a receipt? You could be charged twice." "Oh, no," replied the American, "if God wills I will be back in a week. You can give me a receipt then."
"If God wills," smiled the hotel keeper, "do you still believe in God?" "Why, yes," said the American, "don't you?" "No," said the hotel keeper, "we have given that up long ago."
"Oh," replied the American, "well, on second thought, I believe I'll take the receipt after all!"

The preacher was on the street corner telling the passing crowds about Jesus Christ. A crowd had gathered and was listening intently. Then a hoarse voice spoke up from the back.
" 'Preacher, you've got it all wrong. Atheism is the answer to humanity's problems. People get into trouble and go crazy when they hear about Christianity. Religion is bad for minds and ruins lives. Come on now,—prove to me that Christianity is real, and I'll be quiet.'
Everyone was interested to see what would happen next.
The preacher held up his hand for quiet, and then said this:
"Never did I hear anyone state, 'I was undone and an outcast, but I read Thomas Paine's Age of Reason and now I have been saved from the power of sin.' Never did I hear of one who declared, 'I was in darkness and despair and knew not where to turn, until I read Ingersoll's Lectures, and then found peace of heart and solutions to my problems.'
"Never did I hear an atheist telling that his atheism had been the means by which he had been set free from the bondage of liquor. Never did I learn of anyone who conquered hard drugs by renouncing faith in God.
"But I have heard many testify that, when as hopeless and helpless sinners, they had turned in their great need to the Son of God and cast themselves upon Him for forgiveness and enabling power to overcome sin—they were given peace of heart and victory over enslaving sin!"
Then, turning to the atheist, he said: "Who starts the orphanages, the city missions, and the work among the poor? It is the Christians. Who owns and operates the taverns, and manufactures the liquor sold in them? It is the atheists. Who risk their lives to help poor people in mission fields all over the world? It is the Christians. Who runs the abortion mills and the houses of prostitution? It is the atheists. Who are the most solid, kindly, industrious people in the nation? It is the Christians. Who operates the gambling halls and the crime syndicates? It is the atheists.
"Who are the swindlers, bank robbers, and embezzlers? It is the atheists. Who helps men put away their sins, live to bless others, and prepares men for death and eternity? It is the Christians."

EVOLUTION COULD NOT DO THIS
One research scientist, *T.A. McMahon, worked out the formula for the general size and height of trees. The mathematical formula goes something like this: "The diameter of trees will vary with height raised to the 3/2 power; that is the length times the square root of the length." That is surely a lot for a simple-minded tree, without any brains to keep track of. Here is more of the formula: "The mean height trees obtain is only about 25 percent of that which they could obtain and still not buckle. In other words, trees are designed with a safety factor of about four." Someone very intelligent did the designing. We should not expect that the trees went to college, took math, and figured all that out.
End of Utterly Impossible article

Humans Formed From Dust Used with permission from http://www.anointed-one.net
The Bible says humans were formed from dust:
The Lord God formed the man from the dust of the ground and breathed into his nostrils the breath of life and the man became a living being. Genesis 2:7 written 1400 b.c. By the sweat of your brow you will eat your food until you return to the ground since from it you were taken. For dust you are and to dust you will return. Genesis 3:19 written 1400 b.c. Remember that you molded me

I Was Blind But Now I See   Evolution - Creation

like clay. Will you now turn me to dust again? Job 10:9 written 1900 b.c. He knows how we are formed, he remembers that we are dust. Psalm 103:14 written 1000 b.c. All go to the same place. All come from dust and to dust all return. Ecclesiastes 3:20 written 900 b.c.

For many years, scientists scoffed at the notion these verses could be accurate. Could dust actually be responsible for constructing the complex elements and molecules that make up a human being? After a century of scientific examination of the human body's components, scientists have discovered the ground contains every element, which is found in the human body. NASA's Ames Research Center also came to the same conclusion. Their research confirmed every element in the human body can be found within the soil.

End of info at anointed-one

Elements in the human body.
    Did you realize 99% of your body, by weight, is composed of only six chemical elements? Carbon is the basis for organic life, so you probably knew that element. You know your body is mainly water, so you would guess hydrogen and oxygen are major elements. Can you name the others? Here's a list that shows the elements, gives their mass percent in the human body, and tells you what role they play in your biochemistry.
    The Elements are:
1. Oxygen, 65% of Body Weight. Oxygen is present in water and other compounds. Oxygen is necessary for respiration. You will find this element in the lungs, since about 20% of the air you breathe is oxygen.
2. Carbon, 18.6% of Body Weight. Carbon is found in every organic molecule in the body. Carbon is ingested in food that is eaten and breathed in as a component of air. It is found in the lungs as a waste product of respiration, carbon dioxide.
3. Hydrogen, 9.7% of Body Weight. Hydrogen is a component of the water molecules in the body, as well as most other compounds.
4. Nitrogen, 3.2% of Body Weight. Nitrogen is a component of proteins, nucleic acids, and other organic compounds. Nitrogen gas is found in the lungs, since most of the air you breathe consists of this element.
5. Calcium, 1.8% of Body Weight. Calcium is a major component of the skeletal system. It is found in bones and teeth. Calcium is also found in the nervous system, muscles, and the blood where it is integral in proper membrane function, conducting nerve impulses, regulating muscle contractions, and blood clotting.
6. Phosphorus, 1.0% of Body Weight. Phosphorus is found in the nucleus of every cell. Phosphorus is part of nucleic acids, energy compounds, and phosphate buffers. The element is incorporated into the bones, combines with other elements including iron, potassium, sodium, magnesium and calcium. It is necessary for sexual function and reproduction, muscle growth, and to supply nutrients to the nerves.
7. Potassium, 0.4% of Body Weight. Potassium primarily is found in the muscles and nerves as an ion. Potassium is important for membrane function, nerve impulses, and muscle contractions. Potassium cations are found in cellular cytoplasm. The electrolyte helps to attract oxygen and remove toxins from the tissues.
8. Sodium, 0.2% of Body Weight. Sodium is important for proper nerve and muscle function. It is excreted in perspiration.
9. Chlorine, 0.2% of Body Weight. Chlorine aids in cellular absorption of water. It is the major anion in body fluids. Chlorine is a part of hydrochloric acid, used to digest food. It is involved in proper membrane function.
10. Magnesium, 0.06% of Body Weight. Magnesium is a cofactor for enzymes in the body. Magnesium is needed for strong teeth and bones.
11. Sulfur, 0.04% of Body Weight. Sulfur is a component of many amino acids and proteins.

    Elements found in soil are potassium, calcium, magnesium and phosphorus.
    Sodium is the sixth most abundant element in The Earth's crust, which contains 2.83% of sodium in all its forms. Sodium is, after chloride, the second most abundant element

dissolved in seawater. The most important sodium salts found in nature are sodium chloride (halite or rock salt), sodium carbonate (trona or soda), sodium borate (borax), sodium nitrate and sodium sulfate. Sodium salts are found in seawater (1.05%), salty lakes, alkaline lakes and mineral spring water.

Oxygen is the most common component of the Earth's crust (49% by mass), the second most common component of the Earth as a whole (28% by mass), the most common component of the world's oceans (86% by mass), and the second most common component of the Earth's atmosphere (20.947% by volume), second to nitrogen.

Carbon. Is carbon found in all organic and inorganic matter? The answer is yes and no. Yes, carbon IS found in all organic matter, but NOT in inorganic matter. Although there are many definitions of "organic," in the scientific disciplines, the basic definition comes from chemistry. In chemistry, organic means chemical compounds with carbon in them. In a more general sense, organic refers to living things. And this is connected to the idea of organic chemistry being based on carbon compounds. Organic (carbon-based) compounds are found in all living things. This is where the term "organic" originally came from. If you are thinking that carbon must be very important to life, you are correct. All life on earth can be thought of as "carbon-based." Just be careful about turning this around backwards. It is true that all living things contain carbon compounds... but the opposite is not true. Just because a certain material is referred to as organic does not mean it is or ever was alive. Without carbon, the basis for life would be impossible. Carbon is found in the Earth's atmosphere dissolved in all water bodies.

Hydrogen. Most of the Earth's hydrogen is in the form of chemical compounds such as hydrocarbons and water. Hydrogen gas is produced by some bacteria and algae and is a natural component of flatus (passing gas, farting if you will).

Nitrogen. Nitrogen is an essential part of amino acids and nucleic acids, both of which are essential to all life. The earth's atmosphere consists of 78 percent nitrogen and is the ultimate source of nitrogen. Nitrogen in the air is the ultimate source of all soil nitrogen.

Sulfur. Sulfur is widely distributed in nature. It is found in many minerals and ores, e.g., iron pyrites, galena, cinnabar, zinc blende, gypsum, barite, and epsom salts and in mineral springs and other waters.

Following Used with Permission from http://www.gnmagazine.org/
QUOTE These evolutionists want to maintain their monopoly on what goes into students' minds and not allow that to even be questioned. Where material undermining evolution has made it into the schools, evolutionists have typically fought in the courts rather than through the democratic process where they might be answerable to parents' wishes.

Does it really matter what you believe?

Two opposing worldviews are pitted against each other in this crucial battle for people's minds. The first argues that human beings are nothing but a cosmic accident, the result of millions of years of random mutations and survival of the fittest. The bottom line here is that we should all look out for number one because this life is all there is.

This despairing outlook sums up the Darwinian worldview. In his weekly *Spectator* column, historian and author Paul Johnson analyzes the results: "Much of the blame lies with Richard Dawkins, head of the Darwinian fundamentalists in this country [Britain], who has (it seems) indissolubly linked Darwin to the more extreme forms of atheism, and projected on to our senses a dismal world in which life has no purpose or meaning and a human being has no more significance than a piece of rock, being subject to the blind process of pitiless, unfeeling, unthinking nature" (Aug. 27, 2005, p. 25).

Indeed, Richard Dawkins has described the universe as being characterized by "no design, no purpose, no evil and no good, nothing but blind pitiless indifference" (*River Out of Eden,* 1995, p. 133). Professor McGrath frankly stated that "evolutionary theory leads inexorably to a godless, purposeless world" (*The Twilight of Atheism,* 2004, p. 108).

Clearly the other much more sensible worldview is firmly based on life having a great divine purpose. In the vernacular it says, *"It's not about me"* —that is, this life is not about us seeking to please the self. Instead, it's meant to be about seeking and following the will of the Creator God.

More specifically, the proper worldview is a Christian one, with our life now to be focused on showing love to God and neighbor, striving to become more like the greatest Man who ever lived, Jesus Christ of Nazareth. He showed us the ultimate example of love in giving His life for us so that we might eventually share the entire universe with Him (see Romans 8:16-23).

The stakes in this battle are high. The opposing worldviews shape our thinking (and that of our children) on everything—who we are, why we are here, where we are going, the root causes of our many problems and how, or whether, they can ever be solved.

Don't go into the battle unarmed. Educate yourself as to which is really true—the theory of evolution or the doctrine of creation, particularly as revealed in the Bible. Continue reading *The Good News* to be sure you understand the issues and what's at stake for both you and your loved ones! *GN*

Why Darwinism lives

When rational people consider the intricacies and perfect balance of nature in the world around them, it should become strikingly obvious to them that the marvelous creation requires a Creator. As King David put it: "The heavens tell of the glory of God. The skies display his marvelous craftsmanship" (Psalm 19:1, New Living Translation).

In light of the lack of physical evidence for Darwinism, and abundant evidence against it, why does Darwinism survive? Why hasn't it been discarded like other empty, inaccurate, failed theories?

The apostle Paul answers this question in Romans 1:20-22: "For since the creation of the world God's invisible qualities—his eternal power and divine nature—have been clearly seen, being understood from what has been made, so that men are without excuse. For although they knew God, they neither glorified him as God nor gave thanks to him, but their thinking became futile and their foolish hearts were darkened. Although they claimed to be wise, they became fools" (New International Version).

This passage tells us that the fundamental reason many people reject the biblical account of creation is that they in fact reject God. Although such people may be intelligent and understand many things, when it comes to acknowledging God their thinking is foolishly unsound. The Bible explains, "The fool has said in his heart, 'There is no God'" (Psalm 14:1; 53:1).

People who believe we are merely a part of the animal kingdom reject an important concept that gives us our unique human identity and destiny. The Holy Scriptures reveal that God created us in His image, the "image of God" (Genesis 1:26-27), and gave us the opportunity to become His children (John 1:12). God calls us to become part of the Kingdom of God, not the animal kingdom. God's purpose and plan for humanity are to give every human being the opportunity to acknowledge Him as Creator (Psalm 14:2) and live forever with Him as members of His family (John 3:15-16; 2 Corinthians 6:18).

The biographies of some proponents of Darwinism freely explain why they reject God: They don't want to be subject to God's laws. They want to be free to do whatever they want to do, even act like animals if they so choose. Such thinking leads to and promotes sexual immorality including homosexuality, envy, murder, strife and hatred of God, just to name a few items of a long list of negative qualities inspired by this kind of perspective

(Romans 1:28-31). By contrast, those who aspire to be children of God strive to practice righteousness (1 John 3:10), which means respecting and living by God's instructions.
Unquote

Reprinted with permission of the United Church of God, an International Association. This article is part of a booklet titled "Does God Exist?" and is not to be sold. It is a free educational service in the public interest. Published by United Church of God, an International Association, P.O. Box 541027, Cincinnati, OH 45254-1027. (c)2001 United Church of God. Visit the United Church of God on the Internet at www.ucg.org and www.gnmagazine.org.

The Haeckel Embryos

Even though this was mentioned earlier in my book, I wanted to give a bit more history on this subject. Ernst Haeckel, a German biologist, created drawings that showed embryos from various animals that were identical to each other in their earliest stages. Biologist Jonathan Wells writes that Haeckel's drawings evoked too much alike. Haeckel was more than once accused of scientific falsification. He had doctored his drawings to make the embryos appear more alike than they really were. When his drawings were compared with the drawing of the actual embryo, it became apparent that his drawings were distorted to support his proevolution ideas.

British embryologist Michael Richardson, along with a team of international experts, conducted a study in 1997 comparing Haeckel's drawings with actual embryos. Haeckel's work proved to be one of the most famous fakes in biology. These drawings of course still appear in many recent textbooks and are presented as fact.

Haeckel's Embryos and Actual Embryos:

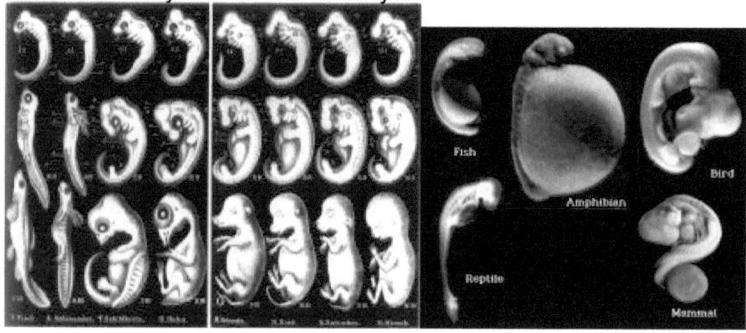

When the author of Evolution Against Science (web link on Resources page) visited the Field Museum in Chicago, they stated the Peppered Moth is still presented as fact even though it had been proven to be fabricated. Even with things that evolutionists no longer believe, they are still being taught as unquestionably true.

This is what the Science Against Evolution said after visiting the Field Museum in Chicago; Quote "Evolutionists keep saying that the theory of evolution is one of the best-established scientific theories there is. They claim there is abundant evidence for evolution. But when placed in a position where they have to put up or shut up, they shut up." Unquote

The Wholly Mammoth

It's most likely that they perished toward the end of the Ice Age, possibly in catastrophic dust storms. Partially digested stomach contents are not proof of a flash freeze, because the elephant's stomach functions as a holding area—a mastodon with preserved stomach contents was found in the western USA, where the ground was not frozen. Read more here online: http://www.answersingenesis.org/tj/v14/i3/mammoth.asp

~~~

"Without God, there is no virtue, because there's no prompting of the conscience. Without God, we're mired in the material, that flat world that tells us only what the senses perceive. Without God, there is a coarsening of the society. And without God, democracy will not and cannot long endure. If we ever forget that we're one nation under God, then we will be a nation gone under."
– Ronald Reagan

~~~

The skeptic asks, "If God created the universe, then who created God?" God, by definition, is the uncreated creator of the universe, so the question, "Who created God?" is illogical. A better question would be, "If the universe needs a cause, then why doesn't God need a cause? And if God doesn't need a cause, why should the universe need a cause?" Everything which has a beginning has a cause. The universe has a beginning; therefore, the universe has a cause. It is important to stress the words "which has a beginning." The universe requires a cause because it had a beginning. God, unlike the universe, had no beginning, so he does not need a cause. Einstein's general relativity shows that time is linked to matter and space. So, time itself would have begun along with matter and space at the beginning of the universe. Since God, by definition, is the creator of the whole universe, he is the creator of time and is independent and outside of time. He is not limited by the time dimension he created, so he has no beginning in time."

Rock from Ica period

In Conclusion

Just because of the Flood, sin didn't simply disappear. Moses was a descendent of Adam and Eve, therefore sin stayed with all. God gave his only begotten son, Jesus, so that our sins can be forgiven. Once you repent your sins, and are ready to receive the Holy Spirit within your heart, truly believe and love God, then you can live forever with him.

Scriptures had been written before the arrival of Jesus, and had to be fulfilled after he was born. Jesus had to be crucified on the cross. He opened a door for us, so that we may go through him to be saved. God has forgiven us of our sins because of Jesus' death, but we must repent those sins and put our faith in the Father, the Son and the Holy Spirit.

You may ask or even wonder – if God is Almighty, then why doesn't he stop all the sin going on now? That's because God wants as many people as possible to be saved before he comes down to judge by way of fire. He wants his Words to reach everyone. As said here in 2 Peters 3:9, The Lord is not slack concerning his promise, as some men count slackness; but is longsuffering to us-ward, not willing that any should perish, but that all should come to repentance. God doesn't want anyone to perish, but everyone to come to repentance. We also must look at several other possibilities to consider explaining all the suffering. Humankind's wrongdoing is also responsible for a great deal of misery and suffering in the world today.

Perhaps we are supposed to continue learning as we have been, from all that God has given us on and around our earth. He created our brains to learn knowledge and wisdom from all things around us. I suspect he wanted us to learn that he created all and hoped that faith in him would happen for many. As for Noah's Ark, maybe God destroyed it, or maybe it's buried deep within our earth or oceans, where we have not been able to go or may never go? He could be testing our faith in him through his Word only, and not by the physical evidence of the Ark. After all, his evidence and proof are quite clear, as you have read in my book.

Just take a look around you, and see all the beauty he created in the trees, the flowers, birds, all living things, including plants. Even your own body, which is very complex, but created just right. Does common sense, logic thinking allow you to believe that your body after millions of years, finally came all together, just fell into place finally to work as a whole? Do you think your body just by chance, without any direction, had evolved by blindness? Did each bird's song take forever and ever to finally get his song just right? Even the complexity of DNA is impossible to have evolved by chance. DNA was designed, what a great creation by God so we may enjoy a variety of different kinds of all on earth.

Even if said, 'perhaps God directed evolution', yes, although he can do whatever he wants, but scripture tells us that this is not how he created. His creation was one of order, not random processes.

Another comment I'd like to make is if you don't believe in God, then you can't believe in Heaven. God and Heaven come as a whole, not in parts. I've heard various people say that of a loved one who passed away, they know they went to Heaven, but they don't believe in God. But if that person does not believe in God, then they cannot believe in Heaven. It doesn't work that way. It's all or none.

God does help those through prayer and the genuine belief and faith you have in Him. He will see that you are provided with your needs. But blind faith, no belief, leads to a life of sinning, no repent. God stated from the beginning what he expects from us and how he wants us to live a good life.

I Was Blind But Now I See   Evolution - Creation

In some caves are drawings from people of long ago. The pictures are pretty straight forward of what it is. So let's take an evolutionist and a five year old to identify these drawings. Say there is a picture of a bird, they both agree. Now a picture of a cat, they both agree. Now we come to another picture, here is the evolutionist's answer, "I don't know." Here is the five year olds answer, "It is a dinosaur." Isn't it funny how the five year old can recognize that man drew a dinosaur? But the evolutionist denied it. Think about this; say someone told you to draw a picture of some animal you have never seen, where would you begin to draw? A person can only draw a picture of something that they have actually seen. And these drawings were done way before dino bones were dug up.

Over 50,000 stones have been found in Peru from the Ica period. These stones contain drawings of dinosaurs we know by name today. There is no rhyme or reason for them to have been "fabricated" so many centuries ago. Furthermore, the fact that they are at least several centuries old is attested to by the oxidation produced by the aging of the minerals covering the incisions of the drawings.

Science continues to uphold knowledge recorded over three thousand years ago from the Bible, as well as agreeing with the Bible's order of creation events.

The Bible isn't just religion or religious, or just a spiritual book. It is a history book written by people who lived it. These books were from scrolls found by the Dead Sea. Why should it be dismissed? Egyptians are part of history, as well as the Old West. They all are a part of history none-the-less. I find it amazing that findings, evidence, proof, all fall back onto the Bible. I'm not going to tell you to believe in Creation or the Bible, but in my humble opinion we didn't evolve, we aren't evolving, nor will we evolve. Evolution is not a proven fact, so it should not be promoted dogmatically. That's why I included links to visit and read. Let your own conclusion help you decide how you arrived. All evidence and proof points to Creation, and evolution doesn't have anything to hold it up, it lacks much, except in the evolutionists mind and imagination.

Nobody has ever proven that the Bible contains any inaccuracies recorded information. Christianity is not really a religion; it is a relationship with God.

It is illogical to think that the different languages and nationalities evolved over millions of years. I just don't see blue-eyed blonde haired apes or redheaded ones with green eyes either. Or any other combination of hair-eye color. I don't see the different languages just happening over time either.

Many know evolution is wrong, but still let you be taught that evolution is the way of the world. As you have read, evolution was disproved ages ago. There is no evidence; there is absolutely no proof. It is all guess work, and nothing to support it, except those who want to control the way you think, and not have your right to believe in what you want, or believe another option to the way the world was created. It's unfortunate that many are not even interested in what the facts are because they could not be contrary to what they already think. People believe what they want to believe. But nowhere online or in a textbook, will it ever say that "particles-to-people evolution is an unsubstantiated *hypothesis* or *conjecture.*"

You do not have to believe in the Bible simply because someone says you are suppose to, that would be blind faith. Blind faith is something you do not need with God. The Bible has been proven to be true.

I don't reject science at all; I only reject Darwin's evolution theory and all that goes with it. There is no evidence, no proof to support that the theory of evolution occurred at any time in the world, the universe or elsewhere.

So... If we evolved from apes, why are they still here? Did they forget to evolve? Even if we are descendents of them, why is it we don't all look exactly alike, why different hair color, eye color, skin color, even different languages? Even if evolutionists say we branched out from apes, there is still confusion where many think we evolved from a

gorilla. Here is what John D. Morris, Ph.D. said on this subject: If Apes Evolved into Humans, Why Do We Still Have Apes? Read file at link provided:
http://marilynoakley.com/apes_humans_question.pdf

The public school systems really need to not teach our children and their children that evolution is a fact of life, when it isn't. I found it amazing how they took this evolution subject and flew off in so many directions with it. In my opinion, they missed a lot of movie calls; the sci-fi channel would never be without material.

This evolution theory has no scientific basis on which to stand. So, it is my hopes that you have learned from reading my book. And in my humble opinion, that evolving or mutation is NOT the answer to how you got here. But how could anyone give Creation science any consideration if they do not know it exists?

Just a thought but when I look at all the different types of dogs, cats, birds, etc. I just don't see blind force creating these at all. It's just hard to fathom that all those evolved or even mutated from one into multiple types. The birds that sing, their song could not have been blind force, just by chance happening. Did each bird make a certain sound or note until something or someone said, you can stop now, sounds perfect? Maybe an ape tapped the bird on the head and gave a thumbs up. God gave me a sense of humor and I'm utilizing it here. Each birds song or sound was designed for that particular bird. It's just not feasible to think that it took millions and millions and millions of years to evolve into what they are today. As for mutation we have yet to find a mutation that *increases genetic information*, even in those rare instances where the mutation confers an advantage.

And as for languages, how could that have evolved? No one is going to tell me, even branching out from a gorilla, that there were different groups of gorillas that decided to invent new languages. I just don't see one group of gorillas in China turning Chinese, or one group in America, turning American, or in Africa, turning African. There were not Mexican gorillas evolving to speak Spanish, there were not Canadian gorillas evolving to speak French.

Whether evolutionists want to say we evolved from ape or branched out, or are ancestors is their opinion. For myself, Creation holds a better account over Evolution. Creation is not just religion; it is an account for our factual history, which it seems certain evolutionists will fight to the bitter end.

It is sad that evolution is so shoved onto our students in public schools, that if they dare challenge it (evolution) or refute it, they may be very well told to hush. Will this tell them they are not free thinkers, no free choice is given, no opinions are to be heard that are against evolution? That they should not examine, question, probe into or analyze Creation or any other option outside of evolution? No student should be made to feel that they are unintelligent because they may question the theory of evolution.

Here's what GN www.gnmagazine.org suggest to deal with believing in God and being taught evolution. Quote

Strategies for overcoming

Now that we have reviewed a few of the fundamental issues of Darwinism and the Bible, what can a student who believes in God do when he takes classes that teach Darwinism? Here are a few strategies: Realize that you are studying a theory. Theories are simply attempts to explain something people don't understand. Darwin didn't understand how human beings came to exist. *The Origin of Species* was his attempt to explain how humans and thousands of perfectly designed species could have come into existence apart from God. You can read his work and study his arguments without agreeing with them. On a test or paper you can write, "Charles Darwin's theory of evolution says . . ." or a similar statement that verifies you know what the teacher or textbook has taught.

I Was Blind But Now I See   Evolution - Creation

It isn't necessary for you to publicly debate teachers or professors who believe in evolution. Through greater experience, they usually have clever, though erroneous, arguments to smooth over the weakness of Darwinism or to make disbelievers in the theory appear ignorant. Remember, the Bible reveals that those who reject God are the ones who are truly foolish and ignorant (Psalm 14:1; Romans 1).

If someone genuinely wants to know what you believe on this issue, tell him (1 Peter 3:15). But you don't have to set yourself up for public or private ridicule. Thinking out your strategy in advance can be quite helpful. Often silence is golden.

If you are asked to do additional research on this subject, consider several possibilities. You might want to read Darwin's book or works on modern variations of his theory and draw attention to the areas in which he and others acknowledge flaws in the theory. Another possibility is to write a book report on material written against evolution.

Use this opportunity to strengthen your relationship with God and your convictions that He is your Creator. Compare the ultimate rewards represented by belief in Darwinism with belief in God. According to the former, when you die you're permanently dead, having no hope of living again. With God you have the marvelous opportunity to live forever in His Kingdom. Don't throw away that opportunity just to fit in with what is currently popular in today's culture. *GN*

Reprinted with permission of the United Church of God, an International Association. This article is part of a booklet titled "Does God Exist?" and is not to be sold. It is a free educational service in the public interest. Published by United Church of God, an International Association, P.O. Box 541027, Cincinnati, OH 45254-1027. (c)2001 United Church of God. Visit the United Church of God on the Internet at www.ucg.org and www.gnmagazine.org. Unquote

The evolution theory has infiltrated the very roots of our society because of the teaching and practice of it. It has worked its way up through the classrooms. The denial of God has always been the root cause of every human problem.

I find it amazing how the national leaders in science and science education who support "evolution only" in the classroom are the same ones who support "free inquiry, open-mindedness and impartiality" in the classroom, but do not allow the free inquiry, open-mindedness and impartiality, because it does not match what the students have already been taught about evolution. Isn't it clearly seen that this is an inherent contradiction in encouraging students to think?

It seems the current scientific establishment and education system are not willing to open up discussion for anything other than evolution. Their fear is that the human mind may come to conclusions contrary to evolution.

It's unfortunate that the courts have declared evolution to be science and creation to be religion, when creation science has many facts and information to offer students. Perhaps evolution should be looked at as a religious issue as it hides behind dogma and authority and the unwillingness to engage in an open dialogue. Face it, you need a lot more faith to believe in evolution.

There was a law that forbade the teaching of evolution. It all began with the growing acceptance of Darwin's theory in the early 1900's. It was through the Scopes Monkey Trial that got evolution into the school's education system. A team of scientists and even theologians traveled to Dayton to help the Scopes' defense (although their testimony was not part of the trial, it is recorded in the transcripts) and proclaim that evolution was true and the law should therefore be struck down. Why? It is my understanding that certain people did not want to accept the real Creation, to deny the existence of God and took it upon themselves to see if evolution could be introduced to students, and introduced as fact.

Marilyn Oakley

So, for the last few decades, evolution has been taught and pushed onto society through education and the media. It is still pushed on everyone today. As a result of the Scopes Monkey Trial, creation and all that went with it, you could say in a sense became ridiculed and disbelieved. Because once our society's mind became brainwashed that evolution was an absolute fact - that questioned the Bible and our actual creation.

This is what was being defended by scientists at the Scopes trial at that time from the Hunter's Civic Biology book. Example from that book on human evolution; "If we follow the early history of man upon the earth, we find that at first we must have been a little better than one of the lower animals." "At the present time there exist upon the earth five races or varieties of man ... the highest type of all, the Caucasians, represented by the civilized white inhabitants of Europe and America." Doesn't this statement seem racist? It does to me. And this is what they teach in evolution. Let's teach all to be prejudice. Somehow, I find this totally wrong.

Dr. Terry Mortenson spoke of the Columbine Massacre of 1999. He said, Quote "Eric Harris, 18, had written, *'Sometime in April me and V (Dylan Klebold, 17) will get revenge and will kick natural selection up a few notches.'* April 20, 1998. This event left 13 dead and 25 injured. One of those boys was wearing a t-shirt on that day that said "Natural Selection". Dr. Mortenson went onto say, that we are not blaming evolution for their behavior, but evolution was clearly an influence in their behavior, and they were thinking "Well, it's survival of the fittest." Unquote

Here is how Hitler used the theory of evolution to murder millions of Jews and others. This quoted by Adnan Oktar (author of Atlas of Creation). "In Nazi Germany, the prevalent belief was that 'The weak will perish and the strong will survive.' Hitler completely accepted Darwin's theory to humanity. The prevalent idea was that the Aryan – the German race – was superior. Hitler claimed that other races were not evolved, and that they were merely ape-like beings and he claimed that they had to be eliminated by natural selection. He believed that wars resulted in caused natural selection – that the elimination of "inferior" races was the law of nature and that they would "naturally" die by natural selection. According to Hitler, the German race was the strong race, at the forefront of evolution of humanity. The other races were transitional species between human and apes. Accordingly, this developed race would eliminate the other, undeveloped animal-like beings by massacring them.

"Darwin was the person who influenced Hitler. Hitler was a very ignorant man, a corporal who was influenced by all the books he read. Consequently, because of the scientific shallowness of his time, because of the weakness of technological tools, he thought that Darwin was really correct and he practiced Darwinism within the social system. And the result of this was a terrible massacre." Unquote Adnan Oktar.

Of course you won't learn of how Hitler used the evolution theory on the History channel. Nazism openly proclaimed its dependence on Darwin and that it was right and moral for the strongest race to survive; to have pity for the weak was to defy nature's laws.

Communism as well took evolution to its logical conclusion that if everything just evolved from 'natural law,' then man's opinion, not God's Word, determined what is right and wrong. Regardless of how many must die and Communism's death toll far outranked the Nazis'—probably more than 90 million worldwide.

Evolution was the chief tool used to brainwash communism's masses into 'scientific atheism.' If everything just evolved, then everything is at the whim of the most powerful, and there is no Maker to whom to be answerable. Mao's reign of terror and lies resulted in the deaths of tens of millions. It is no coincidence that his two favorite books were by the evolutionists Darwin and Huxley. Mao said, 'We have so many people we can afford to lose a few.' His successors have since persecuted and killed hundreds of thousands more. Millions died from his forced famine.

I Was Blind But Now I See   Evolution - Creation

According to a report in The Scotsman newspaper on 20 December 2005, Soviet dictator Joseph Stalin wanted to rebuild the Red Army, in the mid-1920s, with Planet-of-the-Apes-style troops by crossing humans with apes. The report claimed that Stalin ordered Russia's top animal-breeding scientist, Ilya Ivanov, to use his skills to produce a super warrior. Stalin is said to have told Ivanov, 'I want a new invincible human being, insensitive to pain, resistant and indifferent about the quality of food they eat.' Experiments were done by artificial insemination to impregnate chimpanzees. Dictator Stalin, a passionate atheist, based this upon his belief in evolution. Ivanov shared his master's belief in evolution. If evolution were true, humans and apes would be closely related. Of course the experiments did not work of trying to breed human and chimpanzee. God made man in the image of God, not in the image of an ape.

A gruesome trade in 'missing link' specimens began with early evolutionary/racist ideas. But this trade really 'took off' with the advent of Darwinism. Perhaps 10,000 dead bodies of Australia's Aboriginal people were shipped to British museums in a frenzied attempt to prove the widespread belief that they were the 'missing link'. US evolutionists were also strongly involved in this flourishing 'industry' of gathering specimens of 'subhumans'. The Smithsonian Institution in Washington holds the remains of 15,000 individuals of various races. It was Darwin, after all, who wrote that the civilized races would inevitably wipe out such lesser-evolved 'savage' ones.

A German evolutionist who was nicknamed the Angel of Black Death, came to Australia asking station owners for Aborigines to be shot for specimens. A missionary in New South Wales was horrified to witness the slaughter by mounted police of a group of dozens of Aboriginal men, women and children. The heads were then boiled down and the 10 best skulls were packed off for overseas.

Darwinist views about the racial inferiority of Aborigines (backed up by biased distortions of the evidence since shown to be false) drastically influenced their treatment.

A similar horror reappeared in the 1930s, when the evolutionary doctrines of Nazism allowed the consciences of hundreds of doctors, scientists, psychiatrists and other officials to be seared as they set up the machinery to help nature eliminate the unfit.

It's unfortunate how people of yesteryear and even today, use this theory to their own thinking, whether they are right or wrong and they make their own rules.

Friedrich Nietzsche (1844-1900). *Nietzsche was a remarkable example of a man who fully adopted Darwinist principles. He wrote books declaring that the way to evolve was to have wars and kill the weaker races, in order to produce a "super race"* (*T. WalterWallbank and *Alastair M. Taylor, Civilization Past and Present, Vol. 2, 1949 ed., p. 274). *Darwin, in Origin of the Species, also said that this needed to happen. The writings of both men were read by German militarists and led to World War I. *Hitler valued both Darwin's and Nietzsche's books. When Hitler killed 6 million Jews, he was only doing what Darwin taught. It is of interest, that a year before he defended *John Scopes' right to teach Darwinism at the Dayton "Monkey Trial," *Clarence Darrow declared in court that the murderous thinking of two young men was caused by their having learned *Nietzsche's vicious Darwinism in the public schools (*W. Brigan, ed., Classified Speeches).

Evolution is still a fantasy, even after a hundred and fifty years, it is not supported by science at all. If it were a fact, one would think after all this time; something significant or hard data would have turned up to support evolution. As I said, even the earth itself does not support evolution.

Evolutionists say fossils are millions and millions of years old, so how is it that these fossils are exactly alike the well-known species of today? There hasn't been any change from the millions of years ago to the same species of today. No evolving in any way. The millions of years is assumed as well. Am I supposed to swallow that the fossils of millions and millions of years ago evolved back into what they *used to be*? And you have already read the conditions needed to fossilize something. There isn't anything in this life, before or even after that will show any evolving ancestor for anything or anyone on this earth.

After the flood, evidence is apparent in the fossil graveyards and aren't as old as millions of years. Life has such a complex design that it could have never become by chance. The theory of evolution is a major hoax and deception in the history of science.

Let me give you what the evolutionists have said about how the dinosaurs have become extinct. These have actually been taught mind you. Okay, their first theory, dinosaurs became extinct because of constipation. Their 2nd theory, dinosaurs became extinct because of air pollution. 3rd, they said, dinosaurs became extinct because of sunburn. Their 4th theory to dinosaurs becoming extinct; was an asteroid/comet impact. Their 5th theory was *a series of comet impact*. Their 6th theory, dinosaurs became extinct now to a gas explosion. And last but not least, dinosaurs have not become extinct at all; they have merely evolved into ... birds. So, we do have dinosaurs today, but they are birds. So, all you parakeet owners and other birds you may have, are actually dinosaurs. But, you can bet, this isn't their last theory on how dinosaurs became extinct or maybe it will? They still do not have a clue. Somehow, I'll never see the bird the same way again ... it's a dinosaur!

As Dr. Monty White Quotes "As you can see, the evolutionists theory keeps getting rewritten and rewritten and rewritten and rewritten and on and on." Unquote

I can see that there is no end to their theories.

Dr. Monty White also said, Quote "But, in the Bible It Is Written. There is no rewritten and rewritten and so forth.

"Evolutionists are ignorant; they just simply do not want to see the Design in anything. Here's how evolutionists contradict again. In the 1950's, they said we had continuous evolution, no beginning, no end, and always the same. Then much later, it went to the Big Bang theory, which has, a beginning now, that our universe came into existence not only from nothing but also from nowhere." Unquote

Evolution teaches: No God, No Rules. No purpose in life. Since we don't know where we came from, we have no idea where we are going. Making one's life meaningless. How sad, what a great message from evolution. As I said, it takes a lot of faith to believe in evolution. At least God gave us a purpose to live and why. Evolution equals materialism, no love, no morals, and no respect – for oneself, for others or anything else.

Also as said by Dr. Monty White, Quote "If you believe everything came from nothing, then you believe NOTHING ...
    1) ... without mind created reason and logic
    2) ... without intelligence created understanding and comprehension
    3) ... without morals created complex ethical codes and legal systems
    4) ... without a conscience created a sense of right and wrong
    5) ... without emotion created art, music, drama, comedy, literature and dance."
Unquote

Evolution = Natural Selection + Mutation or Evolution = Spontaneous generation or Evolution = Big Bang - ALL actually equals the figment of imagination.

'Evolution is a religion'. A Quote by Michael Ruse, Saving Darwinism from the Darwinisms, National Post, B-3, May 13, 2000. *"Evolution is promoted by its practitioners as more than mere science. Evolution is promulgated as an ideology, a secular religion – A full-fledged alternative to Christianity, with meaning and morality. Evolution is a religion. This was true of evolution in the beginning, and it is true of evolution still today."* Unquote

What do evolutionists fear about creation being taught as well as evolution? Why are they against it? Science is science, whether it is creation or evolution. So what is the big deal? I suppose the evolutionist will always insist that some blob just gradually formed itself together without any direction, guidance or what to actually become. That ALL plant life, its various kinds just happened? That ALL animals, its various kinds just happened? As they say, by way of - miracle? A miracle seems to be their only answer. But an answer without faith, an evolutionists outlook is hopelessness, no purpose in life, so why should they even believe in miracles? A miracle is something only God can perform. If they don't believe in God, then how can they believe in miracles? If man has performed a miracle, then it was a God given gift.

How did these cells/molecules decide among themselves to develop into anything? Did these cells have a brain to think on their own? Of course not, the brain came much later in the millions of years it took to evolve. So how is it that these forms came to life from non-life? I have to assume that any evolutionist can paint a pretty picture and throw some words with it, but it still doesn't make it so.

What cells created rules, laws, morals, a conscience, language, sound, song, music, or art? How did those cells decide to stop evolving or mutating? When did it think to start again and evolve some more? If the evolutionists have many facts, proof and evidence, then why take a stand to hide creation science from the public? Why do the people not have a right to hear ALL? Is the education system only unilateral?

In my opinion, it is NOT up to the evolutionists to decide what the people should be taught. That is a violation to a persons constitutional right of the First Amendment – the Freedom of Speech. Creation science has just as much right to be heard as evolution. Shouldn't all aspects of science be explored? Why is it that the people of a whole do not get to vote on this, do not have a say?

Isn't it suppose to be – We the People, By the People and For the People? A taxpayer has no say in the education system?

We are One Nation Under God, and isn't it time to take a stand to exercise our right to liberty and justice for all?

Are you a freethinker, and do you want your freedom of choice, freedom of speech? Then dare explore outside of the theory of evolution. No one should tell you to not be of free will, even God allows you free will, to think and believe what you want. No one has the right to take that away from you. Not one person can take what God has given. God gives you freedom of choice/speech, I think the education system should follow suit, but that is my opinion as well.

I Was Blind But Now I See (Evolution-Creation) became the appropriate title for my book.

~~~

*A good friend of mine said, "First off there's a ton of scientific evidence that supports evolution." My response, "No, there isn't, just because a text book or other sources say so, doesn't mean it is so. Science since Darwin's time has tried to prove it to no avail."

It's amazing how evolution has made such faithful believers of this religion. As Darwin stated that people made a religion of his queries, suggestions and ideas. Evolution has used nothing but fraudulent methods to deceive the public. It is a theory that should actually die out and be forgotten. If a person doesn't want to believe in God or his creation, they don't have to, but evolution isn't a fact either. *The Origin Of Species* is the evolutionist Bible.

Just looking at the fossil record, and of course not having given any thought to it before, it actually makes more sense that a quick burial preserved these fossils instead of layer after layer over millions of years. Especially with the trees being in upright positions. I'm willing to bet, that even in a million years, that an evolutionist could not even explain the upright position to make it believable. Even my eleven year old can reason that it would have decayed long before being covered by layers. But, imagination can never tell what tale may arise as a just-so story disguised as legitimate science.

Marilyn Oakley

As said by Ken Ham of AnswersInGenesis.com
"The church teaches the Bible – but doesn't connect the Bible to the real world. So our children go into the real world and the real world says...Ah, we teach science and that has nothing to do with the Bible. In church we teach Bible and think it has nothing to do with science. So the world separates God from science and churches tend to separate science from God. "

References and Resources:

http://www.answersingenesis.org
http://www.darwinismrefuted.com/index.html
http://www.halos.com
http://www.cosmicfingerprints.com
http://www.dinosaursinthebible.com
http://www.creationists.org/mananddinos.html
http://allaboutcreation.org
http://www.obvioustruths.com/index.php
http://www.clarifyingchristianity.com/index.shtml
http://www.anointed-one.net/home.html
http://www.samsloan.com/creation.htm
http://www.evolution-facts.org
http://www.creationmuseum.org
http://www.icr.org
http://www.alpha7omega.co
http://www.harunyahya.com/
http://www.creationontheweb.com/content/view/21
http://www.discoverynews.us/
http://www.scienceagainstevolution.org/index.htm
http://christiananswers.net/dinosaurs/video.html
http://s8int.com/index.html

If you don't have a computer or Internet access, perhaps visit a friend or family who does, or visit your local library, they will be happy to assist you in using a computer and the Internet.
This book may be purchased at
http://www.lulu.com/marilynoakley

Recommended Material:
Books: *The Antiquity Of Man* by Sir Arthur Keith
Evolution: A Theory In Crisis by Michael Denton
Know Why You Believe by Paul E. Little
Darwin's Black Box by Michael J. Behe

Books From AnswersInGenesis.org:
Creation Scientists Answer Their Critics by Dr. Duane T. Gish
Darwin's Demise by Dr. Joe White & Dr. Nicholas Camninellis (this new book shows that Darwinism is a fraudulent faith masquerading as science and that creationists have solid evidence to support their claims)

DVD's from the AiG site:
Fossil Record
Creation Evolution and Deception
Origin Of The Species

Recent News

Of its kind cats; thus proving that cats descended from the first cats God created, so it is possible to have many different breeds of cats.

Study Traces Cat's Ancestry to Middle East (2007)

Five subspecies of wildcat are distributed across the Old World. They are known as the European wildcat, the Near Eastern wildcat, the Southern African wildcat, the Central Asian wildcat and the Chinese desert cat. Their patterns of DNA fall into five clusters. The DNA of all house cats and fancy cats falls within the Near Eastern wildcat cluster, making clear that this subspecies is their ancestor, Dr. Driscoll and his colleagues said in a report published on the Web site of the journal *Science*. This scenario has been established by Carlos A. Driscoll of the National Cancer Institute and his colleagues. He spent more than six years collecting species of wildcat in places as far apart as Scotland, Israel, Namibia and Mongolia. He then analyzed the DNA of the wildcats and of many house cats and fancy cats. There was no evolution involved.

Weekly News July 2007

Q: Can speciation occur quickly?
A: If the account of Noah's Ark in the Bible is true, then two of every kind of land animal (and seven of some) came off Noah's Ark in the Middle East. For instance, two members of the dog kind walked off the Ark. Then, as the number of dogs increased, eventually the population split up and different groups formed.

As the gene pool was split up, different combinations of genes—inherited from the original dogs—would end up in different groups. Thus, different species would form, such as dingoes, wolves, and so on.

Evolutionists have often insisted that such a process happens slowly, and therefore, the Bible can't be right when it says that the land animals came off the Ark only about 4,300 years ago.

But in the journal *Science*, a report stated:
"These examples say that natural selection can cause a population to change very quickly and hint that speciation could [occur] very quickly ..."
Once again, true operational science confirms the biblical history. The account of Noah's Ark in the Bible fits with real observations, including natural selection and speciation.

Psalms:118:8: It is better to trust in the LORD
than to put confidence in man.

Marilyn Oakley

Our Emotions Our Minds

How did our minds develop over millions of years? How did our emotions, our conscience develop over millions of years? Why is it within us to feel love, compassion, to like or dislike, hate, and hurt? Why is there God's Law? Because of sin, that's why. Sin brought the emotions of hate, killing, harming and hurting of others. God's law wants love and to treat each other good. Even to love your neighbor as yourself and forgive your enemy.

Even if we descended from ape, how did our emotions come along? Animals are animals. Did these animals' traits just suddenly disappear after we turned into a man? Did emotion develop ASAP? Or did these traits take thousands of years to disappear from man once he came about and then the emotions finally hit the brain? Not likely, because God put those emotions there to care and love, to be compassionate, and man became of sin as in the killing or hurting of others, along with lying, stealing and other wrong doings.

Why may you ask did God allow sin? Because God gave us free will. He wanted to see how our free will would affect us. God didn't put the hate in a man's heart; the man himself put the hate there. God does not tempt people to sin. The emotion may have been within the brain, but under our free will, we chose what emotions we want to use. He left it up to each individual the emotion they want to portray. God gave us a variety of emotions, such as laughter, sadness, shock, and surprise, to name a few.

When God created Adam and Eve, they were created as human beings, to either obey or disobey. Remember, people are responsible for sin – not God. If God had made us so we couldn't sin, then we'd no longer be human beings, but merely machines. He could have made us like robots, but we would have ceased to be human.

One may think why would God send anyone to Hell? God doesn't send anyone to Hell, we send ourselves. God has done what all is necessary for us to be forgiven, redeemed, cleansed and made fit for heaven. All that remains is for us to receive this gift. If one refuses it, God has no option but to give one's their choice.

As quoted by Russell M. Grigg, "God endowed man with intellectual ability which was and is far superior to that of any animal. Thus man was given a mind capable of hearing and understanding God's communication with him, emotions capable of responding to God in love and devotion, and a will which enabled him to choose whether or not to obey God. Man was thus equipped, not only to 'love God and obey Him for ever', but also to do God's work on earth — to be His regent and govern the creation in co-operation with his Creator." Unquote

....

I pledge allegiance to the Flag
   of the United States of America,
and to the Republic for which it stands:
   one Nation under God, indivisible,
With Liberty and Justice for all.
   Thus it is that when you Pledge Allegiance to the United States Flag, You:
   *Promise your loyalty to the Flag itself.
   *Promise your loyalty to your own and the other 49 States.
   *Promise your loyalty to the Government that unites us all,
   Recognizing that we are ONE Nation under God,
   That we cannot or should not be divided or alone,
   And understanding the right to Liberty and Justice belongs to ALL of us.

I Was Blind But Now I See   Evolution - Creation
A Bio On Charles Darwin

Bio from book, entitled Evolution Cruncher (www.evolution-facts.org)
Charles Darwin (1809-1882) was born into wealth and able to have a life of ease. As a young child he was home schooled before attending a local school from age nine to sixteen. He took two years of medical school at Edinburgh University, and then dropped out. It was the only scientific training he ever received. Because he spent the time in bars with his friends, he barely passed his courses. Darwin had no particular purpose in life, and his father planned to get him into a nicely paid job as an Anglican minister. Darwin did not object. **But an influential relative got him a position as the unpaid "naturalist" on a ship planning to sail around the world, the *Beagle*. The voyage lasted from December 1831 to October 1836.**
   It is of interest that, after engaging in spiritism, certain men in history have been seized with a deep hatred of God and have then been guided to devise evil teachings, that have destroyed large numbers of people, while others have engaged in warfare which have annihilated millions. In connection with this, we think of such known spiritists as *Sigmund Freud and *Adolf Hitler. **It is not commonly known that *Charles Darwin, while a naturalist aboard the *Beagle*, was initiated into witchcraft in South America by nationals. During horseback travels into the interior, he took part in their ceremonies and, as a result, *something happened to him*.** Upon his return to England, although his health was strangely weakened, he spent the rest of his life working on theories to destroy faith in the Creator.
   After leaving South America, Darwin was on the Galapagos Islands for a few days. While there, he saw some finches which had blown in from South America and adapted to their environment, producing several sub-species. He was certain that this showed cross-species evolution (change into new species). But they were still finches. **This theory about the finches was the primary evidence of evolution he brought back with him to England.** Darwin, *never a scientist* and knowing nothing about the practicalities of genetics, then married his first cousin, which together they had ten children, three had died.
   His book, *Origin of the Species*, was first published in November 1859. The full title, *On the Origin of the Species by Means of Natural Selection or the Preservation of Favoured Races in the Struggle for Life*, reveals the viciousness of the underlying concept; this concept led directly to two of the worst wars in the history of mankind. **In his book, Darwin reasoned from theory to facts, and provided little evidence for what he had to say. Modern evolutionists are ashamed of the book, with its ridiculous arguments.** Darwin's book had what some men wanted: a clear out-in-the-open, current statement in favor of species change. So, in spite of its laughable imperfections, they capitalized on it.
   Here is what you will find in his book: • Darwin would cite authorities that he did not men-fuller edition" would come out later. But, although he wrote other books, try as he may he never could find the proof for his theories. No one since has found it either. • When he did name an authority, it was just an opinion from a letter. Phrases indicating the hypothetical nature of his ideas were frequent: "It might have been," "Maybe," "probably," "it is conceivable that." A favorite of his was: "Let us take an imaginary example." • Darwin would suggest a possibility, and later refer back to it as a fact: "As we have already demonstrated previously." Elsewhere he would suggest a possible series of events and then conclude by assuming that proved the point. • He relied heavily on stories instead of facts. Confusing examples would be given. He would use specious and devious arguments, and spent much time suggesting possible explanations why the facts he needed were not available.

Here is an example of his reasoning: To explain the fossil trans-species gaps, Darwin suggested that *species must have been changing quickly in other parts of the world where men had not yet examined the strata. Later these changed species traveled over to the Western World, to be found in strata there as new species. So species were changing on the other side of the world, and that was why species in the process of change were not found on our side!* **With thinking like this, who needs science? But remember that Charles Darwin never had a day of schooling in the sciences. Here is Darwin's explanation of how one species changes into another:** It is a variation of *Lamarck's theory of inheritance of acquired characteristics (*Nicholas Hutton III, Evidence of Evolution, 1962, p. 138).* Calling it *pangenesis*, Darwin said that an organ affected by the environment would respond by giving off particles that he called *gemmules*. These particles supposedly helped determine hereditary characteristics. The environment would affect an organ; gemmules would drop out of the organ; and the gemmules would travel to the reproductive organs, where they would affect the cells *(*W. Stansfield, Science of Evolution, 1977, p. 38).*

As mentioned earlier, scientists today are ashamed of Darwin's ideas. In his book, Darwin taught that man came from an ape, and that the stronger races would, within a century or two, destroy the weaker ones. (Modern evolutionists claim that man and ape descended from a common ancestor.) **After taking part in the witchcraft ceremonies, not only was his mind affected but his body also.** He developed a chronic and incapacitating illness, and went to his death under a depression he could not shake *(Random House Encyclopedia, 1977, p. 768).*

He frequently commented in private letters that he recognized that there was no evidence for his theory, and that it could destroy the morality of the human race. "Long before the reader has arrived at this part of my work, a crowd of difficulties will have occurred to him. Some of them are so serious that to this day I can hardly reflect on them without in some degree becoming staggered" *(*Charles Darwin, Origin of the Species, 1860, p.178; quoted from Harvard Classics, 1909 ed., Vol. 11).* "Often a cold shudder has run through me, and I have asked myself whether I may have not devoted myself to a phantasy"*(*Charles Darwin, Life and Letters, 1887, Vol. 2, p.229)*

History Of Evolution

*****Charles Lyell** (1797-1875). Like *Charles Darwin, Lyell inherited great wealth and was able to spend his time theorizing. Lyell published his *Principles of Geology* in 1830-1833, and **it became the basis for the modern theory of sedimentary strata,—even though 20th-century discoveries in radiodating, radiocarbon dating, missing strata, and overthrusts (older strata on top of more recent strata) have nullified the theory.** In order to prove his theory, **Lyell was quite willing to misstate the facts.** He learned that Niagara Falls had eroded a seven-mile [11 km] channel from Queenston, Ontario, and that it was eroding at about 3 feet [1 m] a year. So Lyell conveniently changed that to one foot [.3 m] a year, which meant that the falls had been flowing for 35,000 years! But Lyell had not told the truth. Three-foot erosion a year, at its present rate of flow, would only take us back 7000 to 9000 years,—and it would be expected that, just after the Flood, the flow would, for a time, have greatly increased the erosion rate. **Lyell was a close friend of Darwin, and urged him to write his book**, *Origin of the Species.*

*****Alfred Russell Wallace** (1823-1913) is considered to be **the man who developed the theory** which *Darwin published. ***Wallace was deeply involved in spiritism at the time he formulated the theory** in his *Ternate Paper*, which *Darwin, with the help of two friends (*Charles Lyell and *Joseph Hooker), pirated and published under his own name.

I Was Blind But Now I See Evolution - Creation

*Darwin, a wealthy man, thus obtained the royalties which belonged to Wallace, a poverty-ridden theorist. In 1980, *Arnold C. Brackman, in his book, *A Delicate Arrangement*, established that Darwin plagiarized Wallace's material. It was arranged that a paper by Darwin would be read to the Royal Society, in London, while Wallace's was held back until later. Priorities for the ideas thus having been taken care of, Darwin set to work to prepare his book. In 1875, **Wallace came out openly for spiritism and Marxism, another stepchild of Darwinism.** This was Wallace's theory: Species have changed in the past, by which one species descended from another in a manner that we cannot prove today. That is exactly what modern evolution teaches. Yet it has no more evidence supporting the theory than Wallace had in 1858 when he devised the theory while in a fever. In February 1858, while in a delirious fever on the island of Ternate in the Molaccas, **Wallace conceived the idea, *"survival of the fittest,"* as being the method by which species change. But the concept proves nothing.** *The fittest; which one is that?* It is the one that survived longest. *Which one survives longest? The fittest.* **This is reasoning in a circle.** The phrase says nothing about the evolutionary process, much less proving it. In the first edition of his book, Darwin regarded "natural selection" and "survival of the fittest" as different concepts. By the sixth edition of his *Origin of the Species*, he thought they meant the same thing, but that "survival of the fittest" was the more accurate. In a still later book *(Descent of Man, 1871)*, **Darwin ultimately abandoned "natural selection" as a hopeless mechanism and returned to Lamarckism. Even Darwin recognized the theory was falling to pieces.** The supporting evidence just was not there.

The Challenger was a British ship dispatched to find evidence, on the ocean bottom, of evolutionary change. During its 1872-1876 voyage, **it carried on seafloor dredging, but found no fossils developing on the bottom of the ocean. By this time, it was obvious to evolutionists that no fossils were developing on either land or sea, yet they kept quiet about the matter.** Over the years, theories, hoaxes, false claims, and ridicule favoring evolution were spread abroad; but facts refuting it, when found, were kept hidden.

World War I (1917-1918). **Darwinism basically taught that there is no moral code, our ancestors were savage, and civilization only progressed by violence against others.** It therefore led to extreme nationalism, racism, and warfare through Nazism and Fascism. **Evolution was declared to involve "natural selection"; and, in the struggle to survive, the fittest will win out at the expense of their rivals.** *Frederick von Bernhard, a German military officer, wrote a book in 1909 extolling evolution and appealing to Germany to start another war. *Heinrich von Treitsche, a Prussian militarist, loudly called for war by Germany in order to fulfill its "evolutionary destiny" (*Heinrich G. von Treitsche, Politics, Vol. 1, pp. 66-67)*. Their teachings were fully adopted by the German government, and it only waited for a pretext to start the war (*R. Milner, Encyclopedia of Evolution, 1990, p. 59)*.

Along with mutations, the study of fossils and strata ranks as the leading potential evidences supporting evolutionary claims. But no transitional species have been found. Ancient species (aside from the extinct ones) were like those today, except larger, and **strata are generally missing and at times switched—with "younger" strata below "older." Because there is no fossil/strata evidence supporting evolution, the museums display dinosaurs and other extinct animals as proof that evolution has occurred.** But extinction is not an evidence of evolution.

Marilyn Oakley

American Humanist Association (1933). "Humanism" is the modern word for atheism." As soon as it was formed in 1933, **the AHA began working closely with science federations, to promote evolutionary theory, and with the ACLU (American Civil Liberties Union) to provoke legal action in the courts forcing Americans to accept evolutionary beliefs.** Signatories included *Julian Huxley (*T.H. Huxley's grandson), *John Dewey, *Margaret Sanger, *H.J. Muller, *Benjamin Spock, *Erich Froom, and *Carl Rogers (*American Humanist Association, promotional literature).

Discovery of DNA (1953). *Rosiland Franklin took some special photographs which were used in 1953 by *Francis Crick and *James Watson (without giving her credit), to develop the astounding helix model of the DNA molecule. **DNA has crushed the hopes of biological evolutionists, for it provides clear evidence that every species is locked into its own coding pattern. It would be impossible for one species to change into another, since the genes network together so closely.** *It is a combination lock, and it is shut tight.* **Only sub-species variations can occur** (varieties in plants, and breeds in animals). This is done through gene shuffling (*A.I. Oparin, Life: Its Nature, Origin and Development, 1961, p. 31; *Hubert P. Yockey, "A Calculation of Probability of Spontaneous Biogenesis by Information Theory," Journal of Theoretical Biology, Vol. 67, 1977, p. 398). The odds of accidentally producing the correct DNA code in a species or changing it into another viable species are mathematically impossible. This has repeatedly been established. (*J. Leslie, "Cosmology, Probability, and the Need to Explain Life," in Scientific American and Understanding, pp. 53, 64-65; *E. Ambrose, Nature and Origin of the Biological World, 1982, p. 135).

Revolt in France (early 1960s). **A large number of French biologists and taxonomists (species classification experts) rebelled against the chains of the evolutionary creed** and declared that they would continue their research, but would no longer try to prove evolution—which they considered an impossible theory. **Taxonomists who joined the revolt took the name *"cladists"*** (*Z. Litynski, "Should We Burn Darwin?" in Science Digest, Vol. 51, January 1961, p. 61).

Background Radiation (1965). Using a sensitive radio astronomy telescope, *A.A. Penzias and *R.W. Wilson (researchers at Bell Laboratories) **discovered lowenergy microwave radiation coming from outer space. Big Bang theorists immediately claimed that this proved the Big Bang!** They said it was the last part of the explosion. But further research disclosed that **it came from every direction instead of only one; that it was the wrong temperature; and that it was too even.** Even discoveries in the 1990s have failed to show that this radiation is "lumpy" enough (their term) to have produced stars and planets.

The Wistar Institute Symposium (1966). A milestone meeting was the four-day Wistar Institute Symposium, held in Philadelphia in April 1966. **A number of mathematicians, familiar with biological problems, spoke—and clearly refuted neo-Darwinism in several ways.** An important factor was that large computers were by this time able to work out immense calculations—**showing that evolution could not possibly occur, even over a period of billions of years, given the complexities of DNA, protein, the cell, enzymes, and other factors.** *We will cite one example here:* *Murray Eden of MIT explained that life could not begin by "random selection." He noted that, if randomness is removed, only "design" would remain,—and that required purposive planning by an Intelligence. He showed that it would be impossible for even a single ordered pair of genes to be produced by DNA mutations in the bacteria, *E. Coli* (which has very little DNA), with 5 billion years in which to produce it. Eden then showed the mathematical impossibility of protein forming by chance. He also reported on his extensive investigations into genetic data on hemoglobin (red blood cells). Hemoglobin has two chains, called alpha and beta. A minimum of 120 mutations would be required to convert alpha to beta. At least 34 of those changes require changeovers in 2 or 3 nucleotides. Yet, Eden pointed out, if a single nucleotide change occurs through mutation, the result ruins the blood and kills the organism! For more on the Wistar Institute, read the following book: *Paul Moorhead and *Martin Kaplan (eds.), Mathematical Challenges to the Neo-Darwinian Interpretation of Evolution, Wistar Institute Monograph No. 5.

Antelope Springs Tracks (1968). Trilobites are small marine creatures that are now extinct. **Evolutionists tell us that trilobites are one of the most ancient creatures that have ever lived on Planet Earth, and they lived millions of years before there were human beings.** *William J. Meister, Sr., a non-Christian evolutionist, made a hobby of searching for trilobite fossils in the mountains of Utah. On June 1, 1968, he found a human footprint and trilobites in the same rock, and the footprint was stepping on some of the trilobites! The location was Antelope Springs, about 43 miles [69 km] northwest of Delta, Utah. Then, breaking off a large, two-inch thick piece of rock, he hit it on edge with a hammer, and it fell open in his hands. To his great astonishment, **he found on one side the footprint of a human being, with trilobites right in the footprint itself! The other half of the rock slab showed an almost perfect mold of a footprint and fossils. Amazingly, the human was wearing a sandal!** To make a longer story short, **the find was confirmed when scientists came and found more sandaled footprints.** Meister was so stunned that he became a Christian. **This was Cambrian strata, the lowest level of strata in the world; yet it had sandaled human footprints!** *("Discovery of Trilobite Fossils in Shod Footprint of Human in 'Trilobite Beds,' a Cambrian Formation, Antelope, Springs, Utah," in Why Not Creation? 1970, p. 190)*

Chicago Evolution Conference (1980). While the newspapers, popular magazines, and school textbooks emblazoned evolutionary theory as being essentially proven scientifically in so many ways, the evolutionary scientists were discouraged. They knew the truth. The Switzerland, Wistar, and Alpbach meetings had clearly shown that theirs was a losing cause. However, in yet another futile effort, in October 1980, **160 of the world's leading evolutionary scientists met again, this time at the University of Chicago. In brief, it was a verbal explosion.** Facts opposing evolution were presented, and angry retorts and insults were hurled in return. The following month, *Newsweek (November 3, 1980)* reported that **a large majority of evolutionists at the conference agreed that not even the neo-Darwinian mechanism (of mutations working with natural selection) could no longer be regarded as scientifically valid or tenable. Neither the origin nor diversity of living creatures could be explained by evolutionary theory** *(*Roger Lewin, "Evolutionary Theory Under Fire," in Science, November 21, 1980; *G.R. Taylor, Great Evolution Mystery, 1983, p. 55)*. Why is the public still told that evolution is essentially proven and all the scientists believe it,—when both claims are far from the truth?

Five Polls about Evolution (1954). (1) **The general public supports the teaching of creation in public schools**, not just evolution, by a massive majority of 86% to 8% *(AP-NBC News poll)*. (2) A national poll of **attorneys agree** (56% to 26%) and find dual instruction constitutional (63% to 26%, *American Bar Association-commissioned poll)*. (3) A majority of **university students at two secular colleges also agree** (80% at Ohio State, 56% at Oberlin, *Fuerst, Zimmerman*). (4) Two-thirds of **public school board members agree** (67% to 25%, *American School Board Journal poll)*. (5) **A substantial**

Marilyn Oakley

minority of public school teachers favor creation over evolution *(Austin Analytical Consulting poll; source: W.R. Bird, Origin of Species Revisited, 1954, p. 8).*

Soon after *Darwin's book came off the press, Sedgwick, a contemporary, leading British biologist, wrote him. Noting the ridiculous non-scientific "facts" and hypotheses in the book, Sedgwick warned *Darwin that his book was about to open Pandora's box: "Adam Sedgwick, author of the famous Student's Text Book of Zoology, after reading the book, The Origin of Species, expressed his opinion to Darwin in the following words: 'I have read your book with more pain than pleasure. Parts of it I admired greatly, parts I laughed till my sides were almost sore: other parts I read with absolute sorrow because I think them utterly false and grievously mischievous.' "As feared by this great man of science, the evolutionary idea of civilization has grown into a practical method of thought and code of conduct, affecting the reasoning and actions of every part of the human race. Human conduct is modeled on the philosophy that finds current acceptance."—H. Enoch, Evolution or Creation (1986), pp. 144-145.

It is a known fact that the ACLU has advised every state legislature, considering enactment of a law permitting equal time for both views, that the ACLU will give them another full-blown "monkey trial," as they did at Dayton, Tennessee in 1925. The evolutionists never defend their position with scientific facts, for they do not have any. Instead, they use ridicule and lawsuits (Norman Geisler, The Creator and the Courtroom, 1982; Robert Gentry, Creation's Tiny Mystery, 1986).

Is Atheism Against The Law? Used with Permission by anointed-one.net

Quote - Atheism is a lack of belief mentality which rejects the existence of anything supernatural. By default, atheists are also naturalists and evolutionists. They believe there is a natural explanation for all circumstances and nothing has ever occurred that has a supernatural answer. While atheism does not break any state or federal laws, it does break several scientific laws. A scientific law is defined as the observance and recognition of a repeatable process in nature. It is widely accepted as a statement of fact and a universal truth. Scientific laws do not need complex external proofs. They are accepted at face value because they have always been observed to be true. A miracle is an event which is inexplicable by the laws of nature. A miracle contradicts natural, scientific laws and atheists typically scoff at the suggestion that miracles have ever occurred. What scientific laws does atheism break?

The Laws of Conservation

The laws of conservation are basic laws in physics that state which processes can or cannot occur in nature. Each law maintains the total value of the quantity governed by that law (e.g. matter and energy) remains unchanged during physical processes. Conservation laws have the broadest possible application of all laws in physics and are considered to be the most fundamental laws in nature. In 1905, the theory of relativity showed mass was a form of energy and the two laws governing these quantities were combined into a single law conserving the total amount of mass and energy. This law says neither matter nor energy can be created or destroyed. This fact leads to an inescapable question.

If matter and energy cannot be created, how did they originate?
Where did the entire physical universe come from?

Again, it is impossible to create matter and energy through natural methods. However, they do exist, so we find ourselves in a quandary. It would seem to the unbiased either matter and energy made themselves from nothing or a supernatural creator made them. Both answers violate the law of conservation. The fact that matter and energy cannot be created is consistent with the claim in Genesis which says God rested from his work and all he created. This law of science contradicts the notion that matter came from nothing through natural means. Bible believing theists understand the universe was framed by the Word of God and what is seen did not come from things that are visible. God is the one who calls those things that do not exist as though they did.

Why couldn't the universe have always existed?
Because nothing that has a beginning and an end could have always existed.

Today, virtually all scientists accept the Big Bang theory which says the entire universe came into existence at a particular point in time when all of the galaxies, stars and planets were formed. The Law of Entropy says closed systems go from a state of high energy to low energy and from order to disorder. All closed systems, including our universe, disintegrate over time as they decay to a lower order of available energy and organization. Entropy always increases and never decreases in a closed system. All scientific observations confirm everything continues to move towards a greater state of decay and disorder. Because the available energy is being used up and there is no source of new energy, the universe could not have always existed. If the universe has always existed, it would now be uniform in temperature, suffering what is known as heat death. Heat Death occurs when the universe has reached a state of maximum entropy. It is a fact that one day our sun and all stars in the universe will burn out. Electromagnetic radiation will disappear and all matter will lose its vibrational energy. Because the stars cannot burn forever and because they are still currently burning, they could not have always existed because they would have already burned out by now.

Some believe the law of entropy cannot be applied to the universe because they feel the universe is an open system and not a closed one. A closed system is defined as a system in which neither matter nor energy can be exchanged with its surroundings. Matter and energy cannot enter or escape from a closed system. It has boundaries that cannot be crossed. The definition of the word universe is <u>all matter and energy</u>, including the earth, the galaxies and the contents of intergalactic space, regarded as a whole.

If the universe is "all matter and energy", how could it be an open system?
If the universe is everything, how can there be something else out there to provide more matter and energy?
There is not even one generally accepted scientific theory on the origin of matter and energy.

The Law of Biogenesis

This law is composed of two parts. The first part states that living things only come from other living things and not from non-living matter. Life only comes from life. The second part of this law states that when living things procreate, their offspring are the same type of organism they are. This is consistent with the account revealed in Genesis which says all living things reproduce after their own kind. Sharks only come from other sharks, snakes from other snakes, owls from other owls, orange trees from other orange trees, etc. Every living organism alive today is a product of and evidence for biogenesis. Some people feel biogenesis is not a scientific law, but biogenesis is a law because no one has ever documented a single case of non-living matter coming to life in self-replicating form. It is as true today as it has ever been. On the other hand, abiogenesis has been debunked many times over. When someone observes the first example of spontaneous generation which includes self-replicating machinery (DNA and RNA), biogenesis will no longer be a law. Until that time, it remains one.

Marilyn Oakley

If one stretched out a strand of DNA from the oldest and most basic organism known to man, a bacterium, it would be almost 1,000 times longer than the diameter of the bacterium itself. Its DNA pattern is about 4 million blocks long. Where did all of this exquisite information come from? The components of a bacterium are far more complex than any machine mankind has ever made. There is absolutely zero scientific evidence of the existence of any organisms between the supposed event of abiogenesis and bacteria. This is the biggest missing link of all. There is absolutely no evidence any such organism is alive today or was ever alive in the past. Some feel it makes total sense no such fossils exist because the creature would have been made up of parts which do not fossilize well. If this argument was valid, there would not be any fossils of bacteria but there are.

Replication requires the complex machinery of DNA and RNA which are collectively known as the genome. According to evolution, something like the genome could only achieve its utter complexity through replication, cumulative selection and mutation.

How could DNA and RNA evolve from something very rudimentary into their present day intricacy when the organism containing the basic genome would require the more complex, present day DNA and RNA to replicate?

The Gene Emergence Project has sponsored an event called The Origin of Life Prize. They are currently offering 1.35 million dollars to anyone who can offer a credible, verifiable and reproducible explanation of the origin of life. They are by no means a creation science group. Their advisors include biochemists, molecular biologists, biophysicists, information theorists, artificial life and intelligence experts, exo/astrobiologists, mathematicians and origin-of-life researchers in many related fields. The Foundation's main purpose is to encourage interdisciplinary, multi-institutional research projects by theoretical biophysicists and origin-of-life researchers with special focus on the origin of genetic information/instructions/message/recipe in living organisms. They want to know by what mechanism initial genetic code arose in nature. They are requiring full reign be given to the exploration of spontaneously forming complexity and to inanimate systems of self-organization and replication.

There is not even one generally accepted scientific theory on the origin of life.

Scientific Method
The scientific method is held in high esteem by most atheists and it is composed of the following parts...
1) Careful <u>observation</u> of a phenomenon.
2) Formulation of a hypothesis concerning the phenomenon.
3) Experimentation to demonstrate whether the hypothesis is true or false.
4) A conclusion that validates or modifies the hypothesis.

Nobody has ever observed the creation of matter or energy. Nobody has ever observed a molecular cloud collapse or any planet form. Nobody has ever observed abiogenesis. Nobody has ever observed the evolution of any genome. Nobody has ever observed any phylum, class, order or family change.

Evolutionists are excellent at Step 2 - Hypothesizing.
The only problem comes on Steps 1, 3 and 4 - Observation, Experimentation and Validation.

We read about their theories and the conclusions of the failed experiments they performed in an effort to validate their opinions about a phenomenon that has not only never been proven scientifically but has never even been observed.

The definition of a miracle is an event which is <u>inexplicable</u> by the laws of nature. The fact is there are zero generally accepted scientific explanations on these issues. If you want to believe in naturalism it is fine with me but please don't make the erroneous claim that "science" is on your side.

I Was Blind But Now I See   Evolution - Creation

What term is used to describe something you believe to be true but has no empirical evidence?

Faith.

The bottom line is we live in a universe, which completely frustrates any attempt to explain its origin and content by natural processes alone. The best evidence for the possible existence of a supernatural creator lies in the total lack of any scientific evidence in these key areas. Can God be scientifically proven? No, it would be nice but his existence cannot be proven scientifically. The reason is God is supernatural; he exists outside the natural, scientific world. While our scientific tools cannot prove God exists, they do provide us with evidence we can use to determine if there is a better explanation for what has taken place besides the existence of a supernatural creator.

It is interesting how atheists reject any notion of the supernatural because of what they perceive to be a lack of evidence when they could use that same objectivity to reject their naturalistic worldview. Most atheists are not even honest enough to apply the same burden of proof for naturalism that they demand of supernaturalism.

The laws of science falsify the notion that this physical, living world came to be through natural means. These laws provide very credible evidence for the possible existence of a supernatural being. Atheism violates these basic laws of science. Atheism requires not only a tremendous amount of faith but also a belief in miracles. And not only miracles but natural miracles, an oxymoron. Both naturalism and supernaturalism require faith and which one you place your faith in is one of the two most important choices you will ever make.

Unquote anointed-one

I had read this somewhere and wrote it down. It is beautifully written, and I wanted to include it in my book. Thanks to the person who wrote these words.

"You are a unique creation of God and when he made you, he formed you with a hole inside of you only he can fill. It is human nature to try anything and everything in this world to fill that void, but in the end, only the Holy Spirit of God living inside you can do it. He gave you life and he laid down his life for you so you could live with him for eternity. The very least you can do is give him a few minutes of your life to allow him the opportunity to prove himself to you. Forever is a long time to be wrong."

Marilyn Oakley